Ecodevelopment in Tanzania

Bernhard Glaeser

# Ecodevelopment in Tanzania

## An Empirical Contribution on Needs, Self-sufficiency, and Environmentally-sound Agriculture on Peasant Farms

Mouton Publishers
Berlin · New York · Amsterdam

From the International Institute for Environment and Society,
Science Center Berlin

Translated from the German by David Antal



*CIP-Kurztitelaufnahme der Deutschen Bibliothek*

**Glaeser, Bernhard:**
Ecodevelopment in Tanzania : an empir. contribution on needs, self-sufficiency, and environmentally-sound agriculture on peasant farms ; [a publ. of the Internat. Inst. for Environment and Soc.] / Bernhard Glaeser. [Transl. from the German by David Antal]. — Berlin ; New York ; Amsterdam : Mouton, 1984.
Einheitssacht.: Ökoentwicklung in Tanzania «engl.»
ISBN 3-11-009719-2

*Library of Congress Cataloging in Publication Data*

Glaeser, Bernhard.
Ecodevelopment in Tanzania.
"From the International Institute for Environment and Society, Science Center Berlin."
Bibliography: p.
1. Agriculture--Economic aspects--Tanzania.
2. Agriculture ecology--Tanzania. 3. Farms--Tanzania.
4. Peasantry--Tanzania. I. Internationales Institut für Umwelt und Gesellschaft, Wissenschaftszentrum Berlin. II. Title.
HD2128.5.G56 1984 338.1'09678 83-23736
ISBN 3-11-009719-2

 Printing: Druckerei Hildebrand, Berlin. — Binding: Dieter Mikolai, Berlin. — Cover illustration: K. Lothar Hildebrand, Berlin
Printed in Germany.

# Preface

As we enter the early 1980s, the uncertainties of the future are almost stupefying. The 1960s and 1970s were periods of great expectation and promise perhaps best reflected by several global conferences on environment, habitat, water, health, women and energy - to mention only a few. With the passage of time and very few visible results, hope gave way to cynicism and the erosion of confidence has even led to despair about economic recovery, trade, employment, peace, security and, above all, development. While few from the right or the left disputed the use of the word "development", there were all too few who critically appraised all it meant so that today there is still a search for the "magic of yesteryears" which will again bring about development. The apparent contradictions, capitalism vs socialism; science vs tradition; industrialisation vs cultural growth, are with us and will indeed remain with us. While these big issues are hotly debated, there are hundreds of millions of people in the Third World and elsewhere trying to eke out a living.

This work is about yet another dimension of development - namely ecodevelopment. However, because it tends to have a different emphasis from that favoured by conventional development scientists and economists, it has not attracted the attention it deserves. Most developmental activities tend to be sector-orientated, generally with a bias towards the economic aspects. However, in most developing countries, the majority of the people are outside the formal economic sector; people cultivate their own food crops, build their own houses and raise future generations without financial evaluation of their efforts. These activities take place in a specific social and physical environment, and this fact tends to be underplayed. Ecodevelopment, then, is about the environment and resource sustainability, self-reliance, risk minimisation, and even about indigenous technology which could be improved upon.

In an evolutionary sense, perhaps the first step which has been taken in ecodevelopment is a recognition of the realities of development. It differs considerably from the schools which aim to bring about a transformation by aping development which took place in quite a different context and at the price of destroying entire cultures and even peoples. Such attempts at transformation are irrelevant and will not work in Africa, for they fail to consider the realities of development.

It is important to touch upon some aspects of these realities. First and foremost is the simple fact that the majority of peasants in Africa still live in an environment in which natural resources - such as soil, water and human and animal labour - are the major inputs in agricultural production. The basic physical needs such as food and shelter are still produced locally. Therefore, the conservation and sustainability of these resources are of top priority. Secondly, other necessities such as clothing and services, i.e. health, education and transport, are dependent on non-local factors. Nationally, it is these peasants who are also the main producers of the "cash" or export crops, and the terms of trade have generally declined against primary producers as well as being additionally compounded by very low producer prices. In exceptional cases in which producers have done well, prices have fluctuated so erratically that the overall effect has been perverse. Thirdly, peasants have had very little support and are forced to use the head porterage method for transport as well as outdated equipment and tools. Added to this, the infrastructure is so poor and disconnected that costs tend to escalate. Uncertainties in economic terms have been matched in some areas, with even greater climatic variations which have all but eliminated past gains. In this context, poverty, squalor and disease are not incidentals but realities which development assistance has tried to tackle often in a context which is alien to the recipient, and in cases even contradictory to the stated objectives of development. Finally, where conditions are not static, there is a struggle for equilibrium. This study highlights the reality and the struggle.

In the above context Dr. Bernhard Glaeser presents a bold and challenging piece of research carried out in the Usambara mountains of the Tanga region of Tanzania. In this work, theories are limited to the essential and the focus is on the specific and the empirical. In so doing, the author touches on many facts which are simply ignored by planners. As a consequence of a narrow approach in planning, there have been economic/ecological confrontations. Assistance programmes and projects, even when aimed generally at the poor, will be of value and have a chance of bringing about sustained improvement only if the society in which changes are to take place understands and appreciates what is being done on their behalf, and if this same society can guard its own interests. Under these conditions, difficult choices must be made and the poor will still be prepared to participate in the changing process. This also means that Dr. Glaeser has had to look at changes not only in the environmentally specific sense, but also in the societal context.

In the long term, ecological changes brought about partly through a rapid growth in the population will cause environmental stress which will be difficult to manage. The present difficulties experienced by several of the African countries have their origins in plans and decisions which were taken in the decade prior and subsequent to independence. It is therefore very optimistic to assume that the processes of change can be reversed any quicker than the institution of the changes. For instance, agricultural activities which lead to deforestation and subsequently to soil erosion and a decline in productivity, whether it be of maize or tobacco, cannot simply be reversed by stopping the cultivation of these crops. Trees have to decay before the soil can return to its former levels of fertility. Development, then, needs to be examined in a broader perspective and over an extended period of time. This is one of the messages which should emerge from this book.

In the rush of large-scale regional agendas for agricultural development which have recently appeared for Africa, it is refreshing to read an account of a peasant production strategy of a chosen district in Tanzania. Even though the study tries to link its observation from the Lushoto district to the general ecodevelopment theory and policy, this is not done in a doctrinaire fashion; nor is it administered as a political-economic panacea. One has the impression that the author is suggesting, and rightly so, that more studies be undertaken"for implementing the measures is tedious precisely because details play such a great role". Such details are only possible with careful and patient fieldwork and not through "jet" paced safaris. Collaboration between local and foreign institutions is essential and good studies should be followed up.

Dr. Glaeser's book comes at an opportune time: precisely at a phase in which developed countries are becoming disinterested and the poor countries are showing evidence of an erosion of confidence. This coincides with a time in which the sense of history is quietly being forgotten on questions of trade, peace, security and public intervention, and at a time when aid is on the decline.

The study strongly defends self-reliance. Within the framework of ecodevelopment, there is something which the people themselves can do and development assistance should encourage this. At a time when foreign exchange is so scarce in most developing countries, it should be obvious that capital intensive schemes should not be encouraged. This study puts forward just such a case. However, it is scandalous how glibly development agents and agencies have tended to promote capital intensive schemes and how bureaucrats from the recipient countries have become addicted to such assistance. To a great extent, ecodevelopment promotes low capital development.

Since this study was undertaken in the Tanga region of Tanzania, development assistance from the Federal Republic of Germany has greatly increased. One cannot but be impressed by the output of scientific data, for instance, on hydrology and land use. However, an intriguing research question is whether many of the negative aspects of development alluded to in this volume have been arrested or reversed. This could well be a sequel to this book which Dr. Glaeser and his colleagues could undertake with local researchers and institutions from Tanzania.

It is all too easy to forget the author. Trying to balance the empirical with the theoretical is not always easy but Dr. Glaeser has managed to produce a readable and thought-provoking book; it is also to be hoped that the book will generate some interest. In these difficult times it may be all too easy to conform - to write about overvalued currencies, futuristic technologies, romantic notions of traditions and even ideologies, and among researchers there is the trap of mystification. Fortunately, Dr. Glaeser has spared us all of these.

Adolfo Mascarenhas
Institute of Resource Assessment
University of Dar es Salaam
Dar es Salaam
Tanzania

September, 1983.

# Table of contents

PART B. THE EMPIRICAL EXAMINATION OF A CONCEPT

# Acknowledgements

The project "Ecodevelopment in Tanzania" was financed by the Kübel Foundation Bensheim and the Science Center Berlin.

I am indebted to many of my colleagues for their valuable comments and ideas on the manuscripts, including Andreas Büchting (Kleinwanzlebener Saatzucht AG, Einbeck), Meinolf Dierkes (Science Center Berlin), Georg Elwert and Michael Franke (University of Bielefeld), Katrin Gillwald (Science Center Berlin), and Gerd Spittler (University of Freiburg).

I also thank Kevin D. Phillips-Howard for his sensitive and insightful editorial comments during the preparation of the final manuscript.

Special gratitude, however, is extended to the peasants involved in Tanzania, without whose participation and cooperation this project would never have been possible.

Bernhard Glaeser

# Abstract

Using peasants' crop and farm budgets as a basis, this report analyzes the social and economic feasibility of applying ecological production patterns in tropical African highlands. It then investigates the possible ways to increase yields. The conclusions are based on the premise that the desirability of yield increases cannot be determined independently of basic needs and that a peasant's production strategy should be composed of both risk-minimizing (growing subsistence crops) and profit-maximizing (raising cash crops) elements. The report concludes by linking the Tanzanian case study to general ecodevelopment theory and policy.

# List of figures



# List of tables

# Part A

# The systematic and geographical point of departure

# CHAPTER I

## GOALS AND OVERVIEW

Hunger and malnutrition are a fact of life in vast parts of the world, particularly in the so-called underdeveloped countries. At the World Food Conference of 1974, it was estimated that 462 million people were undernourished and that this number would reach 750 million by the year 1985. It was calculated that food production in the coming years would have to increase by 3.6 percent annually in order to maintain food supplies at their 1970 levels and to prevent mass starvation. With an average annual increase of 2.8 percent between 1970 and 1976, however, agricultural production has not been able to keep pace with population growth. In the "most seriously affected countries" (MSAC), which account for 46 percent of the entire population in the developing world, the annual increase of grain production between 1961 and 1970 actually declined from 2.5 percent to 2.1 percent and is now lagging behind the rate of population growth (Pretzell, 1977).

Yet measures to increase yield often hit economic and ecological limits. One reason is that capital intensive chemotechnical methods of cultivation are, in many cases, too expensive. Another is that they progressively destroy the fertility and structure of the soil and exhaust the land's natural resources, which can be restored only with great difficulty.

For these reasons, there is a necessity to identify agricultural methods that are appropriate for the conditions of the tropical environment and that reconcile economic and ecological demands. The goal of the present study is to ascertain the social and economic conditions governing the introduction of ecologically oriented agricultural production and to evaluate how useful they would be for the society of a developing country, in this case, Tanzania. The area of the investigation included the Usambara mountains in the northeastern part of the country. The field work was conducted over several months in the spring and autumn of 1976.[1] Pilot studies had been conducted in 1974 and 1975 (Egger and Glaeser, 1975).

This report will be presented in the following manner. After clarifying the concepts of ecodevelopment and of appropriate agricultural technology (chapter 2), basic information about the socioeconomic background of the Tanga region is compiled from secondary sources. Information on the investigation area of Shashui was gathered by means of a survey of all households (chapter 3). The report then concentrates on three main topics: needs, the economic situation on peasants' farms, and approaches to an environmentally sound increase in agricultural yield (chapters 4-6).

---

[1] Professor K. Egger (Heidelberg) and Dr. H. W. Schönmeier (Saarbrücken) also participated in this research project, which was financed by the Kübel Foundation (Bensheim) within the Lushoto Integrated Development Project (LIDEP). The Community Development Trust Fund (CDTF) supported the project as the Tanzanian partner organization.

The possibilities for increasing income by improving agricultural yield shape the perspective that guided this investigation. The opinions of the peasants were sought, but in order to assess and integrate them into policy recommendations, the proposed agroecological changes were tested which concentrated at first primarily on erosion control. Because the time available on site was short, the economic evaluation had to be limited essentially to the input factors and to the acceptance of measures by the peasants and the official extension services (chapter 6).

The precondition for any recommendation and consideration of strategy, however, is the de facto analysis of agricultural production on small farms. Chapter 5 presents typical calculations of crop and farm budgets for the most common crops grown for home consumption and for the market. Their classification according to the proportion of land devoted to each shows the peasants' actual production strategies.

Family needs, in turn, determine the production decisions underlying these strategies. The type and rank of primary needs were surveyed in one questionnaire distributed according to a village quota system. The needs were classified by breaking them down into labor and leisure activities (chapter 4).

In this way, a succession of needs emerged ranging from those concerning the actual mode of production to those involving alternative strategies for production and recommendations for farming measures, particularly agroecological ones. The following *results* are to be summarized roughly here: The need structure is still shaped very much by the cyclical lack of food--caused by drought, erosion of slopes, and crop failure. Accordingly, the peasants follow the production strategy that places absolute priority on growing food for home consumption, using only what remains of production reserves to grow market crops for satisfying other primary physiological needs like clothing or improvements in shelter.

In spite of insufficient time for the survey period, peasants and bureaucrats alike were interested in the package of ecological methods for preventing erosion, preserving soil fertility, and effecting long term, optimal increases in yield. The suggested improvements are not intended as a substitute for, but as an improvement in, traditional indigenous production processes and can be useful for both subsistence and cash crops. An optimal planning strategy--one that is conscious of needs--must thus combine risk-minimizing and profit-maximizing elements.

As a consequence of these results, the transferability of the Tanzanian case to other African regions is discussed on the basis of concrete experiences (chapter 7). The Tanzanian case is then integrated into the general ecodevelopment strategy represented by the United Nations Environmental Programme (UNEP), and its implications for theory and policy are examined (chapters 8 and 9).

The goal of "ecodevelopment" and of the New World Economic Order is a change in the international distribution of income that many feel to be unjust. But, whereas the measures taken by the New World Economic Order apply pressure by changing distribution policy in the system of international trade, ecodevelopment aims directly at production and producers, trying at that level to initiate changes by encouraging a sense of self-reliance on indiginous human and natural resources, even if it means a temporary decline in world trade.[1] It can be said, then, that the New World Economic Order is in fact more effective politically, but that ecodevelopment is the more fundamental concept theoretically.

Within the realm of ecodevelopment's policy and strategy falls ecological farming, which advocates practices that make it possible to preserve and utilize the soil's biological potential more effectively. This reduces the need for fertilizers, insecticides, and herbicides, which saves energy and ultimately foreign currency as well. These environmentally sound agricultural techniques are among appropriate technologies designed for the social, economic, and biophysical circumstances prevailing in many developing countries, but are also practiced in industrial countries under different conditions.

---

[1] According to Heckstra (1981, p. 353) the New World Economic Order lacks an ecological foundation. He prefers adoption of the World Conservation Strategy, which is working towards an "international ecological order" integrating conservation with development.

# CHAPTER II

# ECODEVELOPMENT AND APPROPRIATE TECHNOLOGY

## Ecodevelopment

Ecodevelopment policy is still largely pursued on a well-worn path. Macroeconomic measures, especially the encouragement of exports, are supposed to swell the Gross National Product, which must exceed population growth in order to eradicate poverty. This development policy has always failed because of poverty itself, which has not disappeared and which has resulted mainly from the uneven distribution usually associated with it. Only income-related security and social security will reorient traditional societies, whose well-being and security were long reliant on large families. Indeed, since the initiation of international efforts toward development after the Second World War, the gulf between rich and poor nations has steadily widened.

A modified strategy for development, then, begins with the needs of the people. The elimination of poverty is the pre-condition for economic development, not its result. The priority no longer lies in substituting exports and amassing foreign exchange, but in spurring employment in the domestic economy and in ensuring the broadest possible distribution of income. Employment includes self employment, at the subsistence level as well. The peasant's family farm represents a socioeconomically stable unit whose destruction would be detrimental. It can, however, be retained in the larger cooperative form of organization; it can be integrated and preserved. Broad distribution of income means equal treatment, both horizontally and vertically. It means neither regional nor class-specific privileges. Such a concept is not utopian

> in that the means are available to satisfy basic needs: it is a question of distributing them more equitably. Hunger and malnutrition are indeed due to the fact that the poor are deprived of the means either to produce or to purchase their food, the socio-economic mechanisms being so organized as to ensure that the lion's share goes to the rich and the powerful. The satisfaction of the need for food and its production cannot, therefore, be set apart from a transformation of political and socio-economic structures, . . . (What Now?, 1975:30)
>
> Whether in food, habitat, health or education, it is not the absolute scarcity of resources which explains poverty in the Third World, but rather their distribution, traditional mechanisms fostering inequality having been aggravated by an indiscriminate imitation of patterns of the industrialized societies. (What Now?, 1975:29)

This means that

> Another development, geared to the satisfaction of needs, can only be envisaged if this goal is made the very core of the whole endeavour. The production of goods and services as well as its planning and technology must be subjected to it. (What Now?, 1975:29)

In this context, the concept of ecodevelopment diverges somewhat from previous approaches:

> If it is possible for a highly productive society to provide a certain minimum to all its members without abandoning the 'logic of gross product' the question is very different in the Third World. This is one of the reasons behind the search for another development. It is impossible to succeed in satisfying needs that will go on growing and to avoid sacrificing present or future generations without replacing the 'trickle-down' effect by the 'trickle-up' effect--. . . (What Now?, 1975:43)

This concept of need-oriented development can be linked with and considered an integral part of ecodevelopment's programmatic approach to development, which reconciles ecology and economy (Glaeser, 1977). The term "ecodevelopment" was first coined by Maurice Strong, the first Executive Director of UNEP, and was introduced during the first meeting of its Administrative Board in July 1973 in Geneva. Ecodevelopment embraces the rational use of natural raw materials, the application of appropriate technology, and the use of social and political forms of organization that take the existing ecosystem and local culture-specific patterns of experience into account. The goal is to pursue economic development that relies for the most part on indigenous human and natural resources and that strives to satisfy the needs of the population, most of all the basic needs of the poor.

The planning of ecodevelopment projects is shaped by the natural restrictions of the biophysical environment, by the economic limitations posed by the efficiency of, for example, peasant subsistence farms, and by the sociocultural and politico-administrative scope of the extended family or the village community within a selected, geographically definable area. If, for instance, the development of a region, a river basin, or a highland is planned, then a series of such limits must be kept in mind. The size and shape of the target area will be determined by the ability of the planners to harmonize the shift of ecological, economic, and cultural factors. In any case, the populations involved are to be included in the planning process; they should actually be the ones to make and execute the plans.

"Ecodevelopment" thus advocates an alternative concept of development as policy and strategy. The policy of ecodevelopment contains the following elements:

- Needs: satisfaction of basic needs, particularly those of the poorest sections of the population
- Self-reliance: greater use of indigenous human and natural resources
- Environmental compatibility: long-term preservation of the ecological basis of economic development

Ecodevelopment as a strategy seeks to include the following factors:

- Ecological: the introduction of environmentally sound and energy saving farming methods
- Microeconomic: achievement of a sufficient level of individual income, be it in monetary or subsistence terms
- Macroeconomic: the improvement of the employment situation by creating new jobs
- Social: the compatibility with specific local needs
- Cultural: the compatibility with tradition and sets of values
- Political: the inclusion of the population groups involved by having them participate in the project.

Ecologically appropriate methods of production constitute the agrotechnical foundation of ecodevelopment. The principles of ecofarming (see below) can be summarized as follows:

- orderly and diverse production
- recycling of matter and nutrients

- biological erosion control

The application of these principles helps the ecological production pattern maintain and increase the soil's fertility, which is based on available nutrient content, texture, structure, organic matter content, existence of soil flora and fauna, and adequate supplies of air and water. This active compatibility with the environment can help make a contribution to the long-term preservation of the productive basis of economic development. The contrast between ecological and conventional agricultural methods shows that the application of the aforementioned principles can in no way be taken for granted. The high-yielding crop variety and fertilizer technology, which has caught on in developing countries as the "green revolution," encourages overcropping that exhausts the organic and the available nutrient stores of the soil--agriculture's capital resource par excellence--while pursuing short-term maximization of yield and profit. In contrast, ecological production techniques specialize in knowledgeable manipulation of ecological processes and can thereby promote "self-reliance" on indigenous resources. The green revolution technology requires an incomparably higher level of anthropogenic energy in addition to the sun's radiation, a dependence that is satisfied in developing as well as industrialized countries largely by oil imports.

It is true that agriculture in industrialized countries accounts for "only" two to three percent of total energy consumption. The figure climbs to eight percent if the processing and distribution of crops is included. But when calculated for the world's entire population, this order of magnitude corresponds today to a volume of energy of $20 \cdot 10^{15}$ kilocalories for the food sector or forty percent of the total amount of the world population's thermal energy actually converted now. Since the world population will probably double by the turn of the millenium and since growing agricultural products is becoming more energy-intensive, the search for ways to save energy and preserve soil fertility seems totally appropriate (Glaeser, 1978).

Taking the example of the corn crop in the United States, it can be shown that the yield per unit area more than doubled between 1945 and 1970. At the same time, though, the use of nitrogen fertilizers increased sixteenfold (Pimentel et al., 1973), the use of pesticides accelerated from 0 to 22,000 kilocalories per acre. Mechanization (including fuel consumption) and commercial fertilizer devour the most energy --41 percent and 38 percent of total energy consumption, respectively. The energy rate, the ratio of energy output (in food) over energy input (not including solar energy) dropped from 3.7 in 1945 to 2.8 in 1970. The energy rate worsens considerably if the processing industry, farm storage, and refinement are also included in the account. The total expenditure of energy for the food system of the United States from the point of sowing crops to their consumption rose from $685 \cdot 12^{12}$ kilocalories to $2{,}172 \cdot 10^{12}$ in 1970--an increase of more than threefold. Since the ratio between food consumption and population growth remained stable, however, the ratio between consumed food energy and gross energy supply fell from about 1:5 to half that at 1:10 (Steinhart and Steinhart, 1974).

The relative waste of energy in agriculture is illustrated by the significantly lower expenditure of energy in developing countries. How much the introduction of the green revolution technology is changing the circumstances in the developing world, too, however, is shown by the Chinese example. Whereas, vegetable and rice production in Hong Kong still had an energy rate of 24.4 in the 1930s the rate had fallen to an average of 1.3 by 1971, with intensive cultivation that had already taken place in some areas reaching only 0.13 (Newcombe, 1975).

Energy-saving measures are contained in the methods of ecological agriculture. A comparison of sixteen farms being managed according to ecological principles and sixteen being managed according to conventional methods, all in the cornbelt of the United States (Lockeretz, 1975), demonstrated that the same or minimally smaller yields resulted on the ecologically-run farms with an expenditure of energy only one-third that of their counterparts run under conventional conditions. The main reason for this astounding difference seems to lie in conventional agriculture's comparatively low level of nutritional exploitation and to its very great nutrient losses. In addition, the greater intensity of ecofarming certainly play a role.

## The Problematic Nature of Appropriate Technology

The term agricultural technology is often understood to mean "exclusively the use of agricultural tools and machines" (Tschiersch, 1975:8). Tools and machines, then, are employed essentially in connection with manual work, animal tractive power, and motorization.

This very narrow view does not take into consideration the knowledge that must be available to use each tool, knowledge that links the tool to an ends-means context that merely "exists" into a "useable instrument." Using this limited terminology would be like trying to reduce computer technology to hardware.

Instead, there is a more preferable version of technology whose broader linguistic and semantic conceptual range includes the informational component. If *téchne* means a skill or proficiency, then *lógos* means the concept, the theory, or the knowledge of it. Acoordingly, technology is to be defined as a system of ends and means, with the ends being predetermined externally and the choice of a system of means (mechanical, electronic, informational, etc.) being oriented to those prescribed ends and to the limiting environmental conditions involved.[1]

Every technology is basically appropriate to the extent that it is consistent with the idea behind it and that it corresponds to its essential concept. Inappropriate technology is synonymous with technique that is inconsistent with its purpose. It is just as impossible to alter an

---

[1]The definition of technology as a system of means follows Spencer, Simmel, Spranger, Jaspers, Tondl, and Sachsse, among others. On the conceptual systematics of possible interpretations of technique, see Lenk (1973), pp. 202-5.

element (the degree of mechanization, for instance) without changing the technological system, which consists of information and technique as they relate to environment and to that system's ends. Neither may the ends be prescribed as an inevitable part of the system--which would be a little like adjusting the environment to a predetermined technique--because technique in the interest of labeling the means as an end assumes a quasi ideological character and therefore no longer suffices for the concept of appropriate technology.

For agriculture, this means that dwelling only on the degree to which agricultural tools are mechanized falsifies the results of the debate about technology and its transfer. No ceteris-paribus clause can arise here because the sub-aspects and sub-elements of a technological system for agriculture are determined by each other and by the system as a whole. In the following pages, then, *agricultural technology* is taken to mean *a system of potential production procedures in agriculture (type of farming) that takes into account a socially given end such as the socioeconomic and ecological environment.*

In accordance with the concept of technology, its use in the sense of transfer leads directly to the problematic nature of appropriateness--the "correct" integration of technology into the socioeconomic and ecological environmental system. *Transfer* is supposed to mean the transplantation from one economically, socially, and culturally defined politico-geographic area to another regardless of whether the technology in question is a gift or a sale administered by the state or by private business. The respective interests of the donor and the recipient must be examined, which in the final analysis--leaving aside particular interests within the recipient country--again amounts to the criteria of appropriateness. A transfer that does not conform to these criteria and that instead pursues interests diverging from the concept of appropriateness automatically triggers ideological suspicions. The question of whether self-determination is involved or determination by others is an initial clue.

The problem of appropriateness, which here is to be understood as an internal technological problem of development and only secondarily as a politicoeconomic one, can be explored further by examining the case of agricultural technologies. First, the general categorization of transfered technologies can offer a conceptual approach. Ideally, five basic forms of technology can be differentiated within the criteria of appropriateness:

"- the use of suitable indigenous technology
"- the adoption of unmodified technology imported from the industrialized countries
"- the improvement of existing indigenous technology
"- the adoption of modified--usually simplified--technologies imported from the industrialized countries
"- the development of new appropriate technologies in either the developing or the industrialized countries" (Tschiersch, 1975:34)

The first two forms are politically and culturally isolated technologies that already exist. Their counterparts, the next two forms, are hybrids ("upgrading" and "downgrading"). The last form represents a catchall category for combinative new development.

Just what conditions govern the introduction of new technologies is not always a straightforward question. In view of the difficulties that accompany every innovation, does it not make more sense in the long run to introduce the most up-to-date technology available right from the start so that the population and the experts can be spared the problems of having to readapt after just a few years? Finally, difficulties with introducing new technologies during the innovation periods of appropriate technology are greater and more frequent than with modern technologies, which have higher prestige in any case.[1]

Given such difficulties, what speaks in favor of intermediate, perhaps nonmechanized technologies, provided they are appropriate ones?

- The microeconomic argument: risk associated with uncertain expectations
- The macroeconomic and foreign exchange argument: deficits in the balance of trade
- The independence argument: dependence on imports instead of a large degree of technological self-sufficiency

This is arguing directly from the angle of the disadvantages involved in modern technology, which,

---

[1]To avoid culture-specific presumptions when expressing such considerations, it always seems expedient to make a parenthetical comparison with industrialized countries, where appropriate technology also encounters difficulties, as can be observed with the solar house self-sufficient in energy. On the other side of the coin, there is technical equipment that is barely used in the university institutes and oversized EDP facilities even in industry when data processing by subcontract would be less expensive. On a smaller scale, there is the colorful ultra modern typewriter in the office and the electronic pocket calculator, which is extremely impractical unless it has printout paper.

incidentally, would "normally" be preferred as being more productive and even more easily introduced. Yet is there not also the possibility of arguing directly from the agricultural point of view put forward by alternative technology? Such arguments focus on

- a higher rate of employment in the population;
- a long-range preservation of soil fertility by means of eco-techniques.

The employment argument has meanwhile been countered with the view that it is not valid at all until a certain degree of mechanization has been achieved--and then valid only for paid labor but not for the traditional family farm. This seems a sound contention only from the monetary perspective. The absorption of and commitment to labor does not hinge on whether individually performed work presently assumes the form of cash or of kind. It must be said, however, that income in kind does not give an impulse to the cash economy.

The second technical-technological argument stems from the different types of farms there are:

- cooperative vs. individual farming
- cash crop vs. subsistence crop production
- if cash crop producer: selling on export vs. local markets
- if subsistence crop producer: operating on a marginal vs. surplus basis

The distinction between each is not rigid. The coffee or tea farmer also sows corn or cassava on a different plot. Above all, one should be wary of the myth that the subsistence farmer does not want to sell his crops in the first place. The difference is simply that the surplus of corn is sold on the local market while foreign currency is earned only on the export market.

## Ecofarming: An Environmentally Sound and Economically Appropriate Technology?

Ecological farming, the suitability of which for developing countries is to be investigated under the aforementioned criteria, can be defined as a controlled, many-faceted system of mixed cultivation that combines tree, field, and fodder plants, and special crops such as vegatables and coffee. The circulation of organic matter and nutrients and the flow of energy, maintains production and soil fertility with comparatively few anthropogenic inputs. The high stability of the agricultural ecosystem is guaranteed by using biological instead of mechanical means of erosion control (i.e., less mechanical tillage than is required by high yielding crop varieties and artificial fertilizer technology), by tolerating and even taking advantage of weeds to a certain degree, by selecting pest-resistant plant species, and by foregoing maximum yields (after Egger, 1979).

This system of agriculture has implications for more than the physical environment. The economy and society can also be stabilized by the fact that the pattern of methods described fits into sociocultural traditions and values and relies on existing natural and human resources (know-how and information). By reducing the microeconomic risks and the foreign exchange deficit, this system can also stabilize the domestic supply situation and curb dependence on foreign countries.

A technological system with these kinds of perspectives (and one must naturally check to be certain that they are realistic) is not "intermediate" technology, but technology "appropriate" to the country's needs. Whereas the concept of intermediate technology implies that a better and more developed technology already exists but is not being applied for one reason or another, the concept of appropriate technology is that the technology being used in a specific case really is the best for that situation ecologically, economically, socially, and culturally. From this point of view, small scale technology, or even decentralized technology, can be the best in a specific case. In certain situations though, large-scale technology is simply the best, as in the case of steel production.

Ecological farming is based on small, decentralized units that are not very capital-intensive. In comparison with chemotechnical farming, however, its methods operate on a more complex level of information. Information becomes a production factor and thereby helps to conserve other resources. If this agricultural technology proves itself to the farmers, it is indeed "appropriate" and not merely "intermediate."

Depending on the type of farm and taking into consideration the reduction in the degree of mechanization (downgrading) and the upgrading of traditional techniques, two additional questions for ecological farming arise:

- Which tools does the ecotechnique require?

- To what degree can mechanization be used?

One must determine if an increase in the yield per unit of land (not just of labor) can be achieved with a higher degree of mechanization without losing the long-term advantage of preserving resources afforded by the ecofarming approach. One or more such economically and socially sound solutions must be sought. If the tractor is not economically appropriate because it is too expensive, and if the plow is not ecologically appropriate because it digs too deep a furrow, perhaps the oxen-drawn harrow is the answer.

One must also be able to use the catalogue of criteria derived from the concept of appropriateness to test the ecologists' claim that they are introducing an alternative technological system. The economic yardstick should be especially stressed so as to avoid the impression that ideological goals would be asserted at nearly any price--even that of economic development. Economically, the purpose of ecological farming cannot be to dictate the subsistence level for the rural population. A static idea like that would eliminate any possibility of economic development for ninety percent of those living in rural development areas.

An ecological orientation is not to be pursued for its own sake. Ecological farming is not to be seen as an end in itself but as a means to the long-range goal of socioeconomic development. This also includes the production of surplus in the countryside in order to supply the urban population with food and the accumulation of surplus in the countryside in order to establish the preconditions for a mixed economy featuring a division of labor with rural centers of artisans and light industry. The issue of social change becomes virulent here; excluding it is tantamount to a conservative ideology.

The occasionally justified cultural scepticism questions the positive role of technique, challenges the blind faith in technical progress. On the other hand, technical progress in the neutral sense of making headway cannot be assessed negatively in every case even if it has failed to fulfill the hopes nurtured by some social utopians that it could be a condition sufficient for the establishment of an economically classless society. The pendulum of dialectic interpretation has consequently swung back.

In reality, technical progress has always been accompanied by negative repercussions. Mass production did create prosperity, but its new form of goods also created new injustices and imbalances like the exploitation of man and nature, alienation and poverty, consumer fetishism, and socioeconomically underprivileged groups, particularly in remote parts of the world.

But technology itself is not negative. Its negative impact has resulted from the manner in which it has been applied by the privileged industrial societies. In short, we can, technically speaking, just as well produce alternative solutions, too. That is the purpose of looking at technology's degree of suitability. It is to produce more justice, not less; to better maintain socioeconomic equality and ecological balance. In other words, it is to produce Aristotelian *mesótes*, meaning a median of optimal values between extremes of pessimistic values in the area of human society and nature.

In these terms, the ecological pattern of methods cannot be considered an alternative to the scientific, technical development of agriculture in general, but only as a different road for such development. Ecological agriculture is to be discussed and studied empirically with field experiments within the context of general socioeconomic development and not as wistful "back-to-nature" romanticism. Material progress, including the education needed for it, is a valid goal to strive for. It is just that belated development is trying--and should try--to minimize the mistakes that occur as a joint product in the European-American model of development with its two varieties of ownership-- private and state capitalism.

The development problem is also how the colonial legacy of economic, political, and sociocultural backwardness can be eliminated. Is it possible through further material progress in terms of "another development," of qualitative alternatives? Is it possible by "progressing" to precolonial structures? The answer ought to be found in two formative premises guiding the account below:

- The interest of the developing countries lies in pursuing material development
- Economic development is accelerated by refined ecological techniques, but not by a nonhistorical manner of thinking

The following pages take the policies and strategies of ecodevelopment and the measures advocated by ecological farming and investigate how much they can contribute to "accelerated development." Admittedly, development can no longer be measured by the growth rate of the GNP alone. The economy does play a part inasmuch as small farm production output is a decisive factor among the modalities of earning and distributing income. In addition to social, nonmaterial factors, the degree of need satisfaction depends, in turn, largely on income. In the same manner, yield-increases and more effective satisfaction of basic needs must also be valid criteria for the successful application of environmentally sound agriculture.

Accordingly, the following chapters first introduce the reader to the socioeconomic situation of Shashui, the study area in northeastern Tanzania. They will then focus on the three elements comprising ecodevelopment as they relate empirically to that area:

- the need structure (chapter 4)
- the economic aspects of managing small farm production in terms of self-reliance, the use of indigenous resources (chapter 5)
- the possibilities for increasing yields and profits by using energy-saving agricultural techniques that are compatible with the environment and by coordinating these with some supporting measures (chapter 6)

# CHAPTER III

# THE SOCIOECONOMIC SITUATION

## The Tanga Region

The Tanga region of northeastern Tanzania includes the coastal region bordering on southern Kenya and extends westward to the Usambara mountains. Comprising a total of 26,810 square kilometers, it is subdivided into six administrative districts. One of these districts is Lushoto, which has an area of 3,497 square kilometers and which is largely identical to the western Usambara mountains. The study area of Shashui, was located there. The soil in the Usambara mountains consists of loamy sand and reddish-brown to dark gray clay (laterite). An average annual rainfall of 1,100 mm is possible. The annual farming periods are synchronized with the two rainy seasons. Fifty percent of the rain falls during the long rainy season (*masika*) from mid-March to mid-May. The short rainy season (*vuli*) had been very irregular for some years, usually coming in October or November if at all, but it can begin as late as January. In addition, heavy intermittent showers (*mchoo*) can occur in June and July. The highland areas lie 1,000-2,000 meters above sea level. An overview of some of the social science data relevant for the planning of the study is given below (see Tanga Plan 1975-1980, vol. 2, pp. 13-22, 32-48, 52, 166-167, 166-167).

## Population Density and Structure

The population of the Tanga region increased from 771,060 in 1967 to 972,000 in 1975, an annual growth rate of 2.9 percent. The population of Lushoto was estimated to be 271,000 for 1975 with an annual growth rate of 3.2 percent. The population density was 77 persons/$km^2$, the density of the region was 36 and that for the country as a whole was 17 persons/$km^2$. Children under the age of 15 made up 52 percent of the population; 25 percent of them were of school age. Only 38 percent of the population had earning capacity. Because of the population pressure, males between the ages of 20 and 24 were migrating in particularly high numbers, resulting in a ratio of 100 females in the district to only 86 males.

## Settlement Patterns

One-quarter of the population of Tanga lived in settlements of more than 500 inhabitants while three-quarters, or about 700,000 people, were scattered throughout the region or lived in small villages. The highland areas were more densely populated. To improve services like schools, dispensaries, and access to the water supply and to the road network in the countryside, planned settlements with about 250-350 families (approximately 1,500 inhabitants) have been under construction over the years. A disadvantage with this was that some families had been forcibly resettled, which appeared necessary since new houses had to be built and new fields had to be sown. The political goal was defined with a view to taking the nearly 3,000 hamlets that existed in 1976 and consolidating them into 500-600 villages. The resulting agricultural and economic incentives would be a larger area for cultivation and a new impetus for the Ujamaa movement.

## Employment and Income

After 1961, when Tanzania gained her independence, work plummeted in Tanga until 1971. This was due in part to the diminishing importance of the sisal industry, in part to new labor-saving procedures that were introduced in response to higher wages. The entire number of wage earners was halved from 106,000 to 53,500 during the decade, and those employed in the agricultural sector dropped by two-thirds from 90,000 to 33,000. Both numbers have gradually increased in recent years. Today, only half of all male adults in the region still depend exclusively on agriculture for a livelihood. In Lushoto, however, seventy-six employers hired no more than 3,228 white or blue collar workers in October of 1974.

The average monthly income for a male white-collar worker in Tanzania was T.Shs. 391/- in 1973 and varied from T.Shs. 197/- in the agricultural sector to T.Shs. 966/- for finance managers. In 1974 the mandatory minimum wage was raised 42 percent from T.Shs. 240/- to T.Shs. 340/- per month. In 1973 two-thirds of the white-collar workers earned less than T.Shs. 300/- a month, in 1974 about 400/-. Less than a fifth of the white-collar workers earned more than T.Shs. 500/- a month in 1973. This bottom-heavy feature of the income pyramid will become even more conspicuous in the Tanga region because of the high proportion of low-paid sisal workers.

Farming households in Tanzania had an average annual income of T.Shs. 851/- in 1969. Annual income was higher (T.Shs. 1,234/-) in the northeast of the country, however. Including the nonfarming households, 69 percent earned less than T.Shs. 1,450/- in cash a year, and 81 percent earned less than T.Shs. 2,175/-. In the mid-1970s agricultural income--subsistence and cash income--fluctuated between T.Shs. 1,000/- and 3,000/- a year depending on the geographic and ecological circumstances. In Lushoto, income was estimated to lie between T.Shs. 2,000/- to 3,000/-, of which about 30 to 35 percent was in cash.

In 1974 the average per capita product of Tanga was T.Shs. 800/- adjusted for the sisal export tax. In 1975, it was T.Shs. 700/-. The proportion of this in cash income was T.Shs. 660/- in 1974 and 570/- in 1975. The decline was caused by the unfavorable development in the sisal sector. This illustrates the dependence of the region on sisal but does not affect the real income and the living standard of the population. The per capita product of Lushoto must have been far below the figures cited since the value of the marketed crops, in spite of vegetable production, is comparatively low, and nonagricultural branches of the economy scarcely exist.

## Farm Size and Family Labor

The area of the average farm in Lushoto was estimated to be 0.5 to 2 hectares, with land being the limiting factor of production. Regular family labor on such a farm averaged 2 to 3 persons.

## Social Infrastructure

As a result of the diverse distribution of interests during the colonial period, Tanga is today one of the most developed regions of the country. With 1,360 hospital beds in 1971, Tanga's health service ranked first in a regional comparison. Life expectancy in 1967 was 48 years (national life expectancy: 41 years). Child mortality was 10.6 percent for the same year (the national rate, 14 percent).

Tanga had only 275,000 illiterates over 14 years of age, which was well below the national average in 1976. In 1972, 1,004,000 children entered school in Tanzania, corresponding to 50 percent of the relevant age groups. In Tanga, the percentage was far higher.

## Economic Infrastructure

The road network of the Tanga region totalled 2,059 kilometers in 1971, placing the region first in comparison of absolute length and second in relative surface density. Of this network, 279 kilometers were bitumen and engineered gravel roads. Of the 2,603 kilometers of railroads in Tanzania, 279 kilometers are in Tanga. The port of Tanga is the most important after Dar es Salaam and Mtwara. Handling 10,000 passengers annually, Tanga's airport is the most important after those of Dar es Salaam and Kilimanjaro.

## The Study Area of Shashui

### Location

The Shashui study area is located in the Western Usambaras and encompasses 25-30 square kilometers from the market town of Soni in the north to the elevated forest areas in the south, from the Maweni Rocks that dominate the area in the east to the Vugha border in the west. Thus the investigation area did not include the entire administrative unit of Shashui; a few outlying areas were excluded.

Climatically and geographically, Shashui is fairly representative of the Usambara mountains and, to a certain extent, of the East African Highlands in general. With the tremendous difference in altitude from 1,350 meters above sea level in the north to 1,700 meters above sea level in the south and with the moisture guaranteed at the higher latitudes by the stand of forest that borders on the area, various crops often found in tropical mountainous regions can be cultivated: corn, beans, bananas, cassava, cocoyams, cabbage, tomatoes, carrots, lettuce, coffee, and even tea. Cultivation took place in a mixed economy to meet both subsistence and cash needs.

### Settlement Patterns

The population was settled in villages or, more accurately, small hamlets inhabited sometimes by only one clan, an extended family. The hamlets were located on steep slopes or straddled a mountain ridge. Typically, the open space between the dwellings was composed of a reddish-brown loamy soil. Only at higher altitudes was there a well-developed turf, which was kept short by grazing cattle. Houses were made of mud, which filled the spaces in a rough framework of poles and then dried in the sun. The roofs were covered with straw and leaves. The traditional shape for a dwelling is circular, but almost all houses in the study area were rectangular and included three to four separate rooms. Inhabitants with more money were able to afford a corrugated-iron roof and a cement floor. Both promoted the family's health if not the aesthetics of the landscape.

During the day, a village in the study area is very quiet since the inhabitants are working the fields. A few women remain behind in order to take care of housework. Some children play between the houses. Here and there are tethered cattle or goats feeding on banana leaves, grass, or peelings that have been thrown to them. Chickens peck for their food. Upon arrival, the visitor reports to the Balozi, the ten-cell leader, who is usually the head of the clan. If one has the necessary government permission, further negotiations present no problems. The reason for the visit is explained and an appointment is arranged for the desired interview.

The study area of Shashui was comprised of thirty-two such clusters, the smallest of which (Bagamoyo) had two families totalling sixteen members. The largest (Mzungu) had 378 inhabitants divided among sixty-two families. The size of the population in each hamlet was normally 100-200 people. From a socioeconomic perspective, it seemed reasonable to divide the investigation area's thirty-two hamlets into five subareas according to the favorability of each's location vis-a-vis the market town of Soni:

Subarea 1: location near a town

Subarea 2: medium distance from a town

Subarea 3: peripheral areas of Shashui near a town (Maweni Rocks)

Subarea 4: far outside a town in the lower-lying areas (Vugha border)

Subarea 5: far outside a town in the mountainous region (extending to the forest area).

## Population Density and Structure

The population of the study area was surveyed completely by means of a general household inquiry, which could be conducted relatively quickly with the help of contact persons from the respective hamlets.[1] With nearly 4,000 inhabitants living in the 25-30 square kilometers of this area, there was an approximate population density of 130-160/km$^2$. Just for comparison, the population density in the entire district of Lushoto was estimated at 77 inhabitants per square kilometer in 1975 that of Soni division at almost 300 (projection of the <u>Tanga Plan 1975-1980</u>, vol. 2, p. 38). As part of the Soni division, then, Shashui was one of the

---

[1]Conducted together with H. W. Schönmeier (see appendix "General Household Inquiry" with H. W. Schönmeier, 1977).

very densely populated areas. This becomes even clearer when the number of inhabitants is considered in relation to the estimated area of cultivation. This brings the population density per square kilometer to 434.

The 3,939 inhabitants were divided among 595 families (household leaders), which made an average family of 6 to 7 people. Variations between the different subareas were small (see table 3.1).

TABLE 3.1

FAMILY SIZE IN SHASHUI

| Sub-area | Number of Inhabitants | Heads of Household | Average Family Size |
|---|---|---|---|
| 1 | 675 | 93 | 7.3 |
| 2 | 1,065 | 169 | 6.3 |
| 3 | 409 | 60 | 6.8 |
| 4 | 1,140 | 170 | 6.7 |
| 5 | 650 | 103 | 6.3 |
| Total | 3,939 | 595 | 6.6 |

## Sources of Income

Eighty-two percent of all households depended exclusively on agriculture for a livelihood. Only fifty-seven leaders of households had additional work, most of which was situated significantly more often in the areas near the central towns of Soni and Vugha (subareas 1 and 4). Logically, the same was true for the forty-three self-employed persons,

TABLE 3.2

EMPLOYMENT OF HEADS OF HOUSEHOLD ACCORDING TO SUBAREA

| Employment status | Absolute Values within the Subareas | | | | | | Percentage Values within the Subareas | | | | | |
|---|---|---|---|---|---|---|---|---|---|---|---|---|
| | 1 | 2 | 3 | 4 | 5 | Shashui | 1 | 2 | 3 | 4 | 5 | Shashui |
| No information | 0 | 2 | 0 | 3 | 0 | 5 | 0 | 1.2 | 0 | 1.3 | 0 | 0.8 |
| Nonagricultural employment | 26 | 10 | 2 | 16 | 3 | 57 | 28.0 | 5.9 | 3.3 | 9.4 | 2.9 | 9.6 |
| Self-employed | 18 | 18 | 0 | 7 | 0 | 43 | 19.4 | 10.7 | 0 | 4.[illegible] | 0 | 7.2 |
| Agriculture only | 49 | 139 | 58 | 144 | 100 | 490 | 52.7 | 82.2 | 96.7 | 84.7 | 97.1 | 82.4 |
| Total number of heads of household | 93 | 169 | 60 | 170 | 103 | 595 | 100.0 | 100.0 | 100.0 | 100.0 | 100.0 | 100.0 |

TABLE 3.3

SIZE OF PRODUCTION UNITS (IN ACRES) ACCORDING TO INFORMATION OBTAINED FROM CONTACT PERSONS

| Subarea | Acres | | | | | | | | | | | | | Number of Plots | Agricultural Land | Average Farm Size |
|---|---|---|---|---|---|---|---|---|---|---|---|---|---|---|---|---|
| | <1 | 1 | 2 | 3 | 4 | 5 | 6 | 7 | 8 | 9 | 10 | 15 | 20 | | | |
| 1 . . . . . . . | 1 | 11 | 26 | 32 | 13 | 3 | 3 | 3 | 1 | - | - | - | - | 93 | 273 | 2.9 |
| 2 . . . . . . . | - | 6 | 49 | 45 | 30 | 10 | 8 | 10 | 6 | - | 1 | 3 | 1 | 169 | 650 | 3.8 |
| 3 . . . . . . . | - | 1 | 8 | 21 | 18 | 6 | 4 | 2 | - | - | - | - | - | 60 | 220 | 3.7 |
| 4 . . . . . . . | - | 11 | 25 | 63 | 37 | 18 | 8 | 3 | 4 | - | 1 | - | - | 170 | 599 | 3.5 |
| 5 . . . . . . . | - | 1 | 16 | 24 | 24 | 19 | 9 | 2 | 2 | 1 | 2 | - | 3 | 103 | 469 | 4.6 |
| Distribution of households on basis of farm size | 1 | 30 | 124 | 185 | 122 | 56 | 32 | 20 | 13 | 1 | 4 | 3 | 4 | 595 | -- | -- |
| Total acres (Shashui study area) | 0 | 30 | 248 | 555 | 488 | 280 | 192 | 140 | 104 | 9 | 40 | 45 | 80 | -- | 2,211 | 3.7 |

NOTE: 1 acre = 0.40467 hectare

most of them businessmen (merchants) or shop owners (see table 3.2).

## Ownership of Land

Land is inherited by the sons from the father, and, because of its scarcity, it is rarely sold. According to information given by the contact persons, the average cultivated area per family was 3.7 acres, or 1.5 ha. (see table 3.3). Since the area was commonly overestimated by a factor ranging from 1.5 to 3, the average size of the family production unit can realistically be assumed to be 1.2 to 2.4 acres (0.5 to 1 ha). As far as distribution is concerned, there seemed to be a tendency towards smaller farm units in the more heavily populated areas near a town and larger units in the mountainous areas far from a town. The location of these units did not necessarily correlate with farm size. The more plots there were, the smaller their average size and the greater the fragmentation of the farm units. As a rule, a family's plots were widely scattered, often lying as far off as the next village. The average size of a plot was less than one acre, with the smallest ones being found near Vugha and in the subarea somewhat removed from the town and the largest in the mountanous regions far from the town. Given the unreliability of the contact persons, however, who tended to estimate every field at one acre, these figures can only be considered approximate[1] (see table 3.4).

---

[1]Fields were measured during the investigation of farm production units. It was found that their size had

TABLE 3.4

AVERAGE FIELD SIZE ACCORDING TO INFORMATION OBTAINED FROM CONTACT PERSONS

| | Subarea | | | | | |
|---|---|---|---|---|---|---|
| | 1 | 2 | 3 | 4 | 5 | Shashui |
| Average farm size | 2.9 | 3.8 | 3.7 | 3.5 | 4.6 | 3.7 |
| Plots per household | 2.9 | 4.5 | 3.7 | 4.1 | 3.9 | 4.0 |
| Average field size | 1.0 | .9 | 1.0 | .8 | 1.2 | .9 |

The figures on farm size are, for all their inaccuracy, still to be examined for the differences in distribution of land. Only then can a preliminary insight be gained into the degree of income distribution (see table 3.5).

As already indicated, the relative figures are presumably more reliable than the absolute ones. It is important, then, that 80 percent of the farms were within the range of size 1 through 8 (actual size about one-third of an acre to three acres) while only twelve farms exceeded this size. In no case was the size of a farm anywhere near that of a large estate.

## Types of Crops

Subsistence crops like corn, beans, and bananas were grown by all farmers. The picture was different for cash

---

indeed been overestimated, but that the feeling for relative size in all likelihood did have some relation to reality.

crops and livestock. Vegetable growing had become the mainstay in most of the study area. Nine of every ten farmers managed a cash income with plots that were usually tiny and located on the fertile floor of the valley. On these plots, cabbage, tomatoes, or lettuce were grown, the crops usually being planted in the short rainy season (October/November). Coffee has joined vegetables as a main market product, for it requires little labor. Three-quarters of all farmers grew coffee. Since 1974, the Tanzania Tea Authority has encouraged tea growing in the higher regions, and this crop has been accepted by the farmers in spite of drought-related failures. There were isolated fruit (mango, passion fruit) and nut trees, the produce from which is eaten by the children.

## Animal Husbandry

Nearly every household owned a few chickens, which foraged for their own food. The eggs were sold (for T.Shs. -/50 each) or were eaten extremely hard boiled. Almost half of the families kept goats or cattle (see table 3.5). There were almost no pigs because the population is largely Moslem.

## Assets

In addition to the number of fields and livestock, the number of dwellings and their furnishings count as indicators of prosperity. Well-to-do features were considered to be a corrugated iron roof and a cement floor. Of 595

TABLE 3.5

CROPS AND ANIMAL HUSBANDRY IN PERCENTAGE OF THE HOUSEHOLDS

| | Subarea 1 | 2 | 3 | 4 | 5 | Shashui |
|---|---|---|---|---|---|---|
| Vegetables | 92.5 | 88.2 | 95.0 | 91.0 | 93.2 | 91.4 |
| Coffee | 55.9 | 65.1 | 93.3 | 77.7 | 91.3 | 74.6 |
| Tea, Wattle tree | 7.5 | 7.7 | 68.3 | .6 | 49.5 | 19.0 |
| Fruit | 0 | 4.7 | 0 | 11.2 | 27.2 | 9.2 |
| Chickens | 95.7 | 99.4 | 83.3 | 97.1 | 98.1 | 96.3 |
| Goats | 34.4 | 43.2 | 55.0 | 37.7 | 72.8 | 46.7 |
| Cows | 39.8 | 39.6 | 66.7 | 30.6 | 69.9 | 45.0 |
| Pigs | 0 | .6 | 0 | 2.4 | 1.9 | 1.2 |

primary houses (father's dwelling) that were completely inventoried in Shashui, 251 had an iron roof, but only 44 a floor of cement. The majority of the houses had four rooms. The average number was 3.8 and fluctuated across the five subareas between 3.7 and 4.0 (see table 3.6).

Of 595 families, 227 (38.2 percent) owned two or more houses. Of these additional dwellings, which are occupied by the other wives of the head of household, only 49 had a corrugated iron roof and 9 a cement floor. The number of wives a man has showed his prosperity, too, for he had to pay a high price for each bride and had to provide them with houses and fields. The number of wives can be considered as a productive investment in the future as well, however, since the labor expended by every wife increased the prosperity of the family. Of 595 male heads of household, of which 11 were no longer living at the time of the study, 421 were married to one wife, 139 to 2 wives, 23 to 3, and 8 to 4 or more wives (see table 3.7).

TABLE 3.6

DWELLINGS AND THEIR FEATURES

| Features | Subarea | | | | | Shashui | |
|---|---|---|---|---|---|---|---|
| | 1 | 2 | 3 | 4 | 5 | Absolute | Percentage |
| Father's Dwelling | | | | | | | |
| Corrugated iron roof | 50 | 73 | 17 | 70 | 41 | 251 | 42.2 |
| Cement floor | 16 | 15 | 0 | 11 | 2 | 44 | 7.4 |
| Average number of rooms | 3.9 | 4.0 | 3.7 | 3.8 | 3.7 | 3.8 | – |
| Additional Dwellings | | | | | | | |
| Number | 42 | 55 | 20 | 70 | 40 | 227 | 38.2 |
| Corrugated iron roof | 17 | 19 | 1 | 5 | 7 | 49 | 8.2 |
| Cement floor | 4 | 5 | 0 | 0 | 0 | 9 | 1.5 |

## Religious Confession

The percentage of polygamous marriages is naturally due to the Moslems, who make up 87 percent of the population in the study area. This majority is not distributed evenly throughout the area, however. The proportion ranges between 100 percent in the mountainous regions far from a town, to 75.9 percent in the lowlands bordering on Vugha, also far from a town (see table 3.8).

TABLE 3.7

NUMBER OF WIVES PER FAMILY

| | Subarea | | | | | Shashui | |
|---|---|---|---|---|---|---|---|
| | 1 | 2 | 3 | 4 | 5 | Number of Households | Number of Wives |
| 1 wife . . . . | 63 | 122 | 38 | 121 | 77 | 421 | 421 |
| 2 wives . . . . | 25 | 41 | 18 | 33 | 22 | 139 | 278 |
| 3 wives . . . . | 2 | 4 | 3 | 11 | 3 | 23 | 69 |
| 4 wives . . . . | 3 | 1 | 1 | 2 | 0 | 7 | 28 |
| 5 wives . . . . | 0 | 0 | 0 | 1 | 0 | 1 | 5 |
| No information . . | 0 | 1 | 0 | 2 | 1 | 4 | - |
| Total households | 93 | 169 | 60 | 170 | 103 | 595 | 801 |

TABLE 3.8

RELIGIOUS CONFESSION OF THE HEADS OF HOUSEHOLD

| Religion | Subarea | | | | | Shashui | |
|---|---|---|---|---|---|---|---|
| | 1 | 2 | 3 | 4 | 5 | Absolute | Percentage |
| Moslem | 85 | 149 | 51 | 129 | 103 | 517 | 86.9 |
| Christian | 8 | 20 | 9 | 41 | 0 | 78 | 13.1 |
| Total | 93 | 169 | 60 | 170 | 103 | 595 | 100.0 |

# Part B

# The empirical examination of a concept

# CHAPTER IV

## SATISFACTION OF NEEDS IN THE SOCIAL AND OCCUPATIONAL CONTEXT

Ecodevelopment begins with those needs perceived as being acute, particularly the so-called physiological primary needs. The decision to focus on the biological needs basic for survival rests on the value judgment that development measures should be introduced first among the neediest social strata and in the least developed regions, usually agricultural regions, in order to mitigate the worst cases of deprivation. This is the approach taken in this study even though the theoretical concept of needs is in no way exclusively derived from, or legitimized by, the notion of deprivation, but also by such positive factors as encouragement, stimulation, and abundance.

For this reason, it is necessary to explore at least the main aspects of the significance that the accepted needs theory has for a need-oriented approach to development policy. Thereafter, the perspective of participative planning will be taken to study what the most urgent desires of the population are, in which cases of deprivation assistance should be given, and which technological combination is most suitable for such a catalogue of demands.

### On the Theory of Basic Needs

Needs theory stems from motivation theory. The discussion of "higher" and "lower" needs theoretically assumes a hierarchy of needs. The best known empirically derived ranking of needs is by Maslow (1954), who bases it on ontogenetic and phylogenetic experiments and clinical experience (see table 4.1). According to Maslow, the nonsatisfaction of the lower needs prevents the formation of higher needs The need perceived or "felt" most acutely at any given time is an active factor determining human behavior. If the lower needs (hunger, for instance) are satisfied, then new, higher needs emerge and dominate the organism. In most societies the closer the needs are to the upper end of the scale, the less they are satisfied. Maslow cites cognitive needs (the desire to know and understand) and aesthetic needs as preconditions for need satisfaction. On the whole, it is to be noted that individually expressed needs cannot be used as the sole basis for long-range planning. Objective data on needs must be obtained.

Since human actions are mediated not only by physiological needs but also by society, Malinowski (1944) assigns to each need in his series the corresponding cultural reaction (see table 4.2).

These perspectives make two things clear: the theoretical uncertainty of needs research about what a human need really is and the corresponding range of definitions for the term "basic need," which extends from physiological to cultural determinants. It is scarcely surprising, then, when even intelligent attempts to define the concept fail and end up sounding more like empty formulae instead. In the <u>Wörterbuch der marxistisch-leninistischen Soziologie</u> (Dictionary of Marxist-Leninist Sociology), needs are "demands

TABLE 4.1

A HIERARCHY OF BASIC NEEDS

| Basic Needs | Manifestations (Specific needs for . . .) |
|---|---|
| Lower needs | |
| Physiological needs | E.g., food, sleep, warmth, sensory perception, exercise |
| Safety needs | Among children, e.g., the desire for a predictable, regular ordered world, something one can rely on |
| Belongingness and love needs | E.g., group affiliation, affectionate relations, love |
| Higher needs | |
| Esteem needs | a) The desire for strength, success, power, ability, independence, freedom<br>b) Esteem from others (status, dominance, recognition, attention) |
| Need for self-actualization | Doing what one is "capable" of doing |

SOURCES: Based on A.H. Maslow, Motivation and Personality (New York: Harper, 1954; rev. ed., New York-Evanston-London: Harper and Row, 1970), pp. 35-58; K. Lederer and R. Mackensen, Gesellschaftliche Bedürfnislagen [Societal Needs](Göttingen: Otto Schwartz and Co., 1975), p. 53.

for maintenance and development of social and individual life which are formed primarily in the process of production and reproduction and which develop historically."[1] More succinctly and elegantly, Lederer (1977) calls for more precise definitions for a necessary and worthwhile "meaning of human life," (For a more detailed discussion of needs theory and application of theory, see Lederer, 1980.)

On the other hand, the theoretical deficiency should

[1] 2nd ed. (1971), s.v. "Bedürfnisse".

TABLE 4.2

BASIC NEEDS AND CULTURAL REACTIONS

| Basic Need | Cultural Reaction |
|---|---|
| Metabolism | Commissariat |
| Reproduction | Kinship |
| Bodily comforts | Shelter |
| Safety | Protection |
| Movement | Activities |
| Growth | Training |
| Health | Hygiene |

SOURCE: B. Malinowski, A Scientific Theory of Culture and Other Essays, (Chapel Hill: University of North Carolina Press, 1944), p. 91.

On the other hand, the theoretical deficiency should not be overrated in the context of empirical investigations since pragmatic conventions have already evolved--specifically in the research on developing countries. The "Latin American World Model" developed by the Fundación Bariloche (Herrera et al., 1976) is to be mentioned in this regard. Mallmann (1975) has described the main theoretical elements in a co-ordinate system (see fig. 4.1).

Like those above, this concept is still general and is intentionally made more precise by stating the goal that basic needs in the Third World are to be satisfied. It asks which path will lead to this goal (the normatively intentional approach) and not, as traditionally done, which goal will be reached on the path already being taken (the extrapolative-theoretical approach). "A value judgement should be made as to which needs the efforts should be directed and in what

FIG. 4.1. Needs and Satisfiers

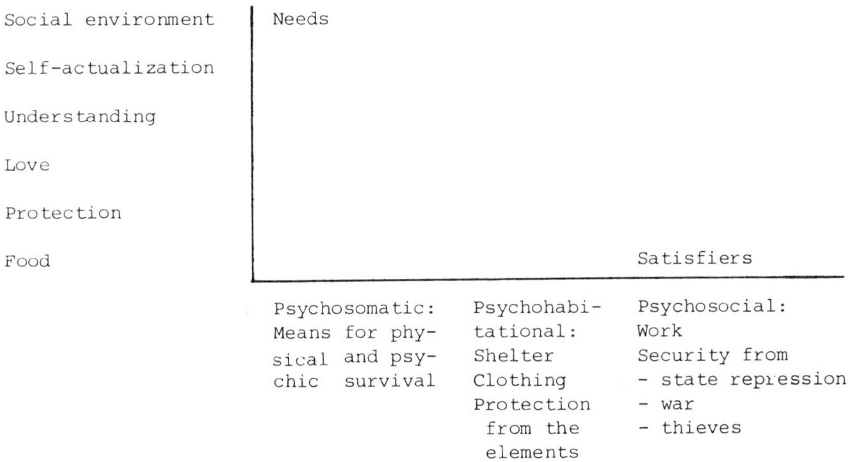


SOURCE: After C. Mallmann, "Die Grundbedürfnisse haben den Vorrang" [Basic Needs Have Priority], Brennpunkte 6 (1975): 17-26.

order of priority" (Mallmann, 1975, p. 20). In terms of this pragmatic restriction, the needs and satisfiers below are stated normatively without further discussing the problem of what a need actually is:

- Food: 3,000 calories and 100g of protein per person daily
- Shelter: 50 $m^2$ of covered space per family
- Education: 10 years of school and 3 years of advanced training
- Full employment

This catalogue of basic needs can be justified pragmatically by the deprivation that very obviously prevails in most developing countries. By the same token, however, one could add such needs as

- clothing
- health

- transportation
- recreation

The assertion that deprivation is endemic implies that the population is living--in economic terms--either at or below the poverty level. Such a statement, whose "objective" physiological criteria are limited at best, is very difficult to operationalize. There is also the "intersubjective" historical situation of those involved as well as the "subjective" cultural prejudices of the outsider. For example, the dichotomy between objective physiological needs and subjective social needs is unsatisfying inasmuch as the concrete qualitative manifestation of all needs is transmitted socially. This is just as true for dietary behavior as it is for the definition of health. Different means of satisfaction are required at different levels of development. In addition, new needs are created as old ones are satisfied. The following need and satisfaction scale has not been derived from theory but has been heuristically accepted and is to provide the basis for further investigation:

- Food
- Clothing
- Shelter
- Health
- Education, training
- Employment, social security
- Transportation, communication
- Recreation, leisure, vacation

Further itemization of the problem area will operationalize the inquiry.

Which needs are listed by the inhabitants themselves? To be sure, current opinion of needs researchers (Maslow, for instance) considers this step irrelevant for long-range planning because of the change of needs mentioned above. Yet, such questions can still prevent planners from neglecting those needs that are the greatest in a given situation (negative criterion). Moreover, there is an interest in comparing the theoretically accepted needs hierarchy to the needs actually expressed at a specific point in time.

What are the characteristics of the income situation that can be externally observed? This involves identifying the more or less objective facts given, specifically, the existing demand and the limited nature of the available satisfiers.

Which needs can be satisfied by the available cash or subsistence income? A two-part question is relevant here. First, which type of demand (need) can be met (satisfied) and to what extent with the available resources? Second, which demand (need) would it be necessary to meet (satisfy)? This is where scientific evaluation is necessary.

## Needs in the Study Area

### Background Information

During his study of 80 households located in the Lushoto District, P. Meyer asked the heads of family what their main expenditures were (Meyer, 1971, pp. 16-17). If it is assumed that the relatively small amount of cash income each had was spent on the most essential items first--goods with the highest capacity for providing satisfaction--then counting the frequency of expenditures can, within certain limits, reveal a hierarchy of satisfied needs (see table 4.3).

TABLE 4.3

FREQUENCY OF EXPENDITURE

| Items | Frequency | |
|---|---|---|
| Food | 25 | |
| Clothing | 20 | |
| Shelter, cooking utensils | 24 | |
| Agricultural investments | | |
| New land | 2 | 8 |
| Cattle | 6 | |
| Radio, bicycle ("luxury goods") | 4 | |
| School fees (education) | 9 | |
| No answer | 27 | |

SOURCE: After P. Meyer, "Vegetable Growing in Lushoto District." Paper prepared for the Lushoto Integrated Development Project (LIDEP), Soni, 1971, pp. 16-17. (Mimeographed.)

In accordance with the theoretical scale of needs, the expenditures for food, shelter, and clothing were listed more often than any other items, with the ranking not necessarily being significant. Even so, expenditures for education placed fourth while health was not mentioned because free medical care already existed. The same has since become true for school fees. Directly following were the expenditures for agricultural means of production. Herein lies the essential discrepancy between this and the hypothetical ranking, in which the need for land and livestock was not included anywhere because the theory had been tailored to the urban population. Of course, it should be considered whether or not the expenditures cited should be organized according to food needs and employment needs. Transportation and recreation were not mentioned by those polled, which likewise is to be explained (though only in part) by the lack of money to spend. Expenditures for radios and bicycles either exceeded the concept of a basic need or, contrarily, fulfilled the need for communication and transport.

In the more qualitative survey conducted by Glaeser in twenty-five production units in the district of Lushoto in 1974, questions on the past use of income were supplemented with some on future use: If you had more money, what would you do with it (Glaeser, 1975, p. 163)? From this dual perspective, needs can, with less hesitation be inferred as a kind of interpretation of a scarcity or desire whose satisfaction has been brought into the realm of possibility (see table 4.4).

While not representative, this survey is consistent with Meyer's results on the frequency of expenditures. The expenditures for food, clothing, and shelter were the ones listed most frequently, with the overwhelmingly high proportion of food expenditures indicating the scarcity of supply caused by the long dry season. New to the catalogue of the responses given to the unstructured-response question was the category of "savings," which could perhaps be subsumed in the category of "social security." Only one of the interviewees possessed an automobile as a means of transportation, and he also used to go on vacation once a year.

Some of the responses about future desires completely inverted the ranking of past needs. Here, long-term investment (house, agriculture) came first, followed by desires for clothing, more and better food, and the accumulation of reserves.

TABLE 4.4

EXPENDITURES AND DESIRES

| Needs in Lushoto | Number of Expenditures | |
|---|---|---|
| | Incurred | Anticipated |
| Food | 9 | 2 |
| Clothing | 5 | 5 |
| Shelter (house, roof, furniture) | 2 | 9 |
| Agricultural investments | | |
| Land, implements | 1 } 1 | 4 } 6 |
| Cattle | - | 2 |
| Automobile, radio ("luxury goods") | 2 | - |
| Savings | 2 | 1 |
| No answer | - | 1 |

SOURCE: After B. Glaeser, "Sozialökonomische Bestandsaufnahme und Problematik in den Usambara-Bergen" [Socioeconomic Inventory and Problems in the Usambara Mountains], in Politische Ökologie der Usambara-Berge in Tanzania [The Political Ecology of the Usambara Mountains in Tanzania], pp. 139-222, by K. Egger and B. Glaeser, Kübel Foundation Series (Bensheim: Kübel Foundation, 1975), p. 163.

Given this discrepancy between expenditures and desires, which hypothesis best reflects the need profile in Lushoto? The following interpretations are offered in answer to this question. Basic physiological needs, which can be satisfied quickly and whose symptoms of scarcity are felt here and now, are satisfied with existing resources first. Expectations that increase with income are quite another matter, for they create basic needs that can be satisfied only over a longer period of time. Both are certainly relevant to planning, and the planner orients himself primarily to the need standard such as the statistically verifiable, objective scarcity and the available resources. He does not look at subjectively expressed desires until the elaboration phase. A "beautiful" house, for example, topped the list of long-range needs in Lushoto. The dream house was no longer to be the traditionally windowless, mud-and-pole structure but a brick one with a cement floor, a leak-proof, corrugated-iron roof, and more rooms with large doors and windows (Glaeser, 1975, p. 164). After establishing the need for and scope of planning, the planning specifications should be set, taking the expressed desires into consideration, and employing the technology best suited to and most appropriate for the particular situation.

In an investigation conducted at the beginning of 1976, Schönmeier (1977, pp. 131-142 and appendix 1.5, pp. 271-272) again questioned seventy-six men on their needs and keyed their answers to a preestablished means hierarchy: What would you do if you had 5,000/1,000/200 shillings at your disposal? The answers varied, with the priority of needs changing according to the sum of money. With 5,000 shillings, agricultural investments (coffee field, poultry farm, wire fence) were listed more frequently than any other category. With 1,000 shillings, priority went to both agricultural investments *and* consumption needs (fields, clothing, food), while purely consumption needs (food) became most important when only 200 shillings were to be made available.

The effect of the change in needs brought on by increased income or by greater satisfaction of the "lower" (immediate, short-term) basic needs emerged as postulated on the basis of this theory. This interpretation is further supported by the fact that "wealth" was cited as a kind of "general need" to which all other needs were like steps along the path of satisfaction. Shashui residents known to be particularly well-off were named as personal models--people who were admired and who were considered worth imitating. Consequently, the question about nonmaterial needs like health elicited no responses worth mentioning. The material desires expressed in this context concerned money, house, car, truck, and tea fields. These answers no longer come under the definition of basic needs, but the result of the survey probably reinforces the hypothesis about the fluid transition of needs into each other, the postulate of a need hierarchy that exists only in the mind and that cannot really be broken down into, say, "basic" or physiological needs and "higher" social needs.

This result also confirms how irrelevant expressed needs are to planning unless they are combined with "objective" data on needs and unless the people affected are themselves engaged in the planning process. To conclude, it is worth noting that at least a partial satisfaction of expressed needs is possible through one's own effort and personal involvement -- in other words, through work. Similarly, work input is also cited as a reason (certainly not the only one) for the wealth of the recognized models.

A range of questions yet to be answered arises from the discourse so far.

Comprehensiveness. Is it possible to make more precise and, above all, more complete statements about the ranking of basic needs in Shashui than researchers have been able to make in the past? How high are nonmaterial needs like education or health ranked compared to material needs?

Production reserves. Apparently, potential production reserves are not yet exhausted even at the present level of technology. It is not the peasant's knowledge or expertise that is lacking; there seems instead to be a discrepancy between knowledge about agricultural techniques and their use. This is the central question to which the following investigation of needs narrows, for it has already been mentioned that the people surveyed believed work input would contribute significantly to the satisfaction of individual needs but that this effort was not always being made. Perhaps this is one reason for the failure of the various development projects in Usambara, a striking example for which is the Usambara Scheme of the 1940s (Mlalo Rehabilitation Scheme), although it was plagued by political difficulties as well (Hozza, 1969).

A few questions relevant to a needs survey do indeed crystalize. Which goals are linked to the personal input of labor? Is perhaps the maximization of profits or harvest that seems so natural to us--and this is added as a hypothesis--forsaken altogether in favor of subsistence farming? Economic rationality, the achievement of a goal with the least expenditure of means, can be imputed to both patterns of behavior. How else might such a goal be harmonized with the desire for wealth if not by considering additional effort unprofitable after a certain point?

Willingness to work and the marginal benefits of labor. The latter supposition leads to an inversion of the question. The goals, the desired goods, are no longer the subject of inquiry; the means required for achieving them are. When does the subjective marginal benefit of labor approach zero?

## Survey: Methodology and Procedure

The methodological problem of needs research in Africa is that need hierarchies created and tested in Europe or the United States cannot in principle be transferred to other cultures. The difficulty in being able to make statements about the ordering of needs in the investigation area of Shashui is that there is no theory of needs for developing countries, a situation that can lead one to ask the wrong questions. Despite this void, an investigation of needs outside the European culture can yield relatively reliable results if

- the survey is prepared by in-depth interviews with prominent personalities and is then conducted in the framework of a group discussion
- the survey results are supplemented by an examination of indicators and objectively-obtained data.

Questionnaire. The above needs hierarchy, beginning with food, clothing, and shelter and continuing through health, education, employment, transportation, communication and recreation is presented in the questionnaire "Basic Needs" (see appendix ) and left open to discussion in the form of a "structured response item" (question 01). A preliminary answer will be sought for the question of whether there are empirical deviations from the theoretical needs scale in the Shashui study area.

In preparatory meetings, experts pointed out that such a ranking was too abstract and too long for non-English reading interviewees to remember. During the group discussions, however, there was keen interest in precisely this item. First, the interviewer explained the question about rank ordering of needs, then read aloud all the available alternatives, afterwards reading them again section by section. During this part of the discussion, the groups elaborated a clear and plain, though not always unanimous, notion of the relative importance of each alternative.

Specifically, the questions about farmland and work status were intended to operationalize the need for employment and social security; the questions about important local means of transportation and appealing travel destinations, the need for transportation and communications; and the questions about free time, the need for recreation, leisure and vacation, according to the social circumstances.

The fact that the questions may have overlapped contrasts with the clear delineation between needs in theory and reveals that the responses this survey elicited are not necessarily reliable despite the assertiveness with which they were given. They must be reexamined. That was the purpose of subsequently paraphrasing the subject, particularly when asking about the needs hierarchy as manifested concretely in labor and leisure behavior. An unstructured-response item about the justification of the chosen priorities (question 02) was to guage understanding and to detect possible "cultural centrism."

Questions 03-08 focused on the ordering of needs in the selection of occupation and the satisfaction of those needs through *work behavior*. Using generally familiar occupations typical for the social strata, the purpose was first to ascertain which occupations were considered attractive and if the shift to the "dream occupation" emerged as a desire. Which motives influenced the desire for the ideal occupation? On the other hand, what background information on social environments was given? The occupations named also represented a hypothetical hierarchy--that of status represented by income and education--rising from the farmer and progressing through the wage earner, the artisan, and the teacher to the self-employed worker and the party and government job. The reason for the desired or actual selection of occupation was, in turn, an indicator of needs, in this case represented by the work situation. This study also attempted to identify the "higher" social basic needs that Maslow postulates. The occupations perceived as best lending themselves to satisfying needs were determined by the process of selection. A rank ordering emerged by counting the preferred occupations, not by regarding the individual responses.

Both of the first two alternatives in question 07 aimed at eliciting the priority of subsistence income versus cash income. The following item on the use of income (question 08) sought to specify these answers and cross-checked the answers given under question 01. The next four choices in question 07 concern state service, steady income, savings, and retirement money differentiating the domain of employment needs and social security needs. The final five choices under question 07 operationalized the social needs for belongingness and love, esteem, and self-actualization. The validity of these choices was restricted because the behavior code of Shambaa forbids public acknowledgement of the desire for admiration, a fact that was discovered during the survey itself.

Need satisfaction according to *leisure behavior* was the subject of questions 10-11. The unstructured-response item on preferred free time activity was intended to avoid the especially acute danger of seeing through European "cultural lenses." In fact, however, the responses all remained within the range of choices given in the structured response format, which, like those concerning work behavior, were to be evaluated positively or negatively without individual weighting. The choices available had to do with the evaluation of individual, group, or communication needs.

Pombe is an extremely popular, slightly alcoholic beverage fermented from corn and bananas with the help of sugar cane leaven. A public endorsement of pombe drinking is, however, tabooed by the party and the government quite apart from the fact that Moslems do not drink anyway. Statements 11.4 and 11.5 were intentional contradictions to the larger question on leisure needs-- whoever is forced to work because of poverty has very little free time. The statement "I work for the party or the government (11.6) meant either honorary political or administrative activity at the village level like teaching or village management (as a ten-cell leader or village secretary) or participation in Ujamaa activities in the broad sense of building roads, irrigating, or clearing forests. The final statement (11.7) focused on the need for training. The "higher" social basic needs for love, safety, esteem, and self-actualization were addressed, in the order of their occurrence, via the idea of better education.

The items about work time spent in the field and the possibilities for increasing it (questions 09 and 12) led to the topic of production reserves in the sense that basic needs determine the production strategy--maximizing crop yield versus minimizing risk--and the desired labor intensity --balancing marginal costs between decreasing marginal productivity and increasing hardship and suffering.

Supplementary items on overpopulation and the scarcity of land (questions 13-15) were designed to furnish information about the reaction of the population specifically to the evaluation of two options of emigration and family planning. This also included state-directed emigration in the form of establishing "planned settlements."

<u>Conducting the survey</u>. The survey was carried out with the help of an interpreter in Kiswahili or Kishambaa in the framework of group discussion with 12 groups from 11 of the 32 hamlets located in the investigation area of Shashui. A

total of 45 persons, which included two groups of women with a total of 13 participants, took part in the discussions. Some of the groups came to unanimous responses while some found themselves in dissent. The questionnaire was prepared by earlier visits to the area and pretested in intensive interviews with highly-placed, educated Shambaa. The ques-iohnaire results were also cross-checked by observation and against objectively-obtained data.

## Results

Needs and their family orientation. Food was clearly at the top of the need ranking system (see table 4.5). It was placed first by 37 of the 45 interviewees. More than half of those interviewees--the absolute majority--chose clothing and shelter as the second and third priority, respectively. This underscores the central role of the classical minimal needs for Shashui. The explanation for the order of priority chosen coincides with the quantitative statement. Most of the explanations were explicitly based only on the three highest needs cited. The explanation, which was often re-peated almost word for word, was "food, clothing, and a house are needed most. Without food you cannot live; without clothing you must run around naked; and, after all, one does need a house for shelter."

The ranking order was less unequivocal from the fourth position on, obtaining only pluralities. This changed again with the last two points, which elicited totally negative responses. An absolute majority among the

TABLE 4.5

NEEDS HIERARCHY

| Need Satisfiers | Hypothetical Ranking | Ranking by Experts | Ranking by 45 Interviewees | Ranking by the Men | Ranking by the Women |
|---|---|---|---|---|---|
| Food | 1 | 1 | 1 | 1 | 1 |
| Clothing | 2 | 2 | 2 | 2 | 2 |
| House | 3 | 3 | 3 | 3 | 3 |
| Health | 4 | - | 5 | 5 | 5 |
| Education | 5 | - | 4 | 4 | 4 |
| Agricultural land | 6 | 5 | 6 | 6 | 4 |
| Employment | 7 | 4 | 8 | 7 | 9 |
| Transportation | 8 | - | 7 | 8 | 7 |
| Travel | 9 | - | - | - | 8 |
| Leisure | 10 | - | - | - | - |

NOTE: The ordering of the surveyed need satisfiers corresponds to the hypothetical hierarchy derived from the literature on the subject and from previous experience. The experts deviated from this ordering in that they acknowledged only half of the available alternatives as important.

interviewees not only rejected leisure time or travel as a need, but considered the very question nonsensical. The comment was "without work, you would have nothing to eat." One of the groups of women interviewed responded with virtually the same words. Incidentally, only one position had negative responses: employment (in terms of earning wages) was rejected on the grounds that such activity would not leave enough time for working in the fields.

Education ranked above health, arable land, and transportation to the market. Medical aid may play a slightly less important role since the medical services in the district are, all in all, very well developed. On the other hand, the relatively great significance attributed to education is instructive. The system ranges from elementary schools to adult education, and the relevance of a good school education for the level of an individual's standard of living, especially for the children, is recognized clearly.

One group of women showed their very pragmatic common sense on this question. Batuli gets better prices for her goods at the market because she has already attended third grade. Besides, school education is important for raising children. Incipient thoughts about emancipation may be apparent in the answer, "How can I find a job? After all, I am a woman and have learned nothing other than farming fields. . . . Well, with an education it would be possible to. . . ."

The significance of the various answers given by women could not be tested because of the small number of women available for interviews. It is simply noted that the activities performed by women in particular have by their very nature received greater attention in the articulation of the priority scale. This is true for hard agricultural labor and care for the home and family, which is effected more in the importance of cultivated land for meeting personal needs. But it is also true for the arduous trips to market, during which heavy loads must often be carried on the head for miles (A. Fleuret, 1976; P. Fleuret, 1976), and the desires this nurtures for means of transportation.

The significance of the need for arable land has to be seen in the light of overpopulation and land scarcity. It is known that many Shambaa go to the "plains"; particularly to Korogwe, in order to plant corn without settling there for this purpose. Only half of the groups interviewed denied that family members migrated to a different area. This indicates a relatively high rate of migration, although precise numbers cannot be quoted (Tanga Plan, 1975-1980, Vol. 2, pp. 43-6). In the present context it is much more interesting to find out if the scarcity of land as a cause for migration also includes the extreme case of being landless and if family planning could be an alternative strategy to migration to the unpopular hot plains of the country (questions 13-15).

Landlessness was denied in 10 of the 12 villages surveyed; it was cautiously confirmed in two. In Vuli there were people with "very little land" who as a result had to farm in the plains. Four persons in Ntuii did not have more than a quarter of a hectare, which was clearly so little that the contact person could not call it a farm.

There were no plans for resettlement in the district of Lushoto during the survey, which was general knowledge among the interviewees (question 15.1).[1]

---

[1] By resettlement schemes is meant the transfer and grouping of villages into larger units that facilitate measures affecting the infrastructure, more efficient use of resources, training programs--in short, all types of development measures and programs. One disadvantage is the scarcity of land that necessarily results in the immediate vicinity of such artificially created large villages.

Family planning is not a realistic development alternative in the rural area studied here. Other possibilities must be pursued instead, especially yield increases that also protect resources and preserve the soil. Eight out of twelve groups rejected family planning altogether. Only four were able to imagine that it would be better if most people had fewer children. As soon as the general point of view got down to the level of personal action, the result became much clearer. Only two participants of one group intended (according to their own statements!) to limit the number of their children to six and eight, respectively. To the European, this scarcely seems like a limitation. But more important than the statements, which cannot be verified anyway, is the justification that one must be in the position to feed and clothe the children properly. The mother must stay in good health despite the strain caused by children.

These answers were given by young men with comparatively high levels of education. Otherwise, the traditional attitude in favor of an unlimited number of children seemed to be stronger in the men than in the women, an attitude reflected in the following interview with women. The women indicated that they acquiesced to the wishes of their husbands in these questions. Naturally, the traditional idea among women, too, is that you should bring as many children into the world as God gives you since even birds, which do not till the land, beget sons.

It was very difficult to conduct a discussion of needs with women or groups of women since a trusting relationship had to be established with the villagers first, especially with the village chief. Only toward the end of the study was it possible to arrange two interviews. Because these two interviews in particular turned out to be quite interesting, one of them is reported below.

The first attempt at a discussion with women ran aground when the chairman of Shashui village did not appear and did not inform the women. It seemed likely that he wanted to obstruct the matter. The second attempt to organize a discussion did succeed. The chairman was summoned from the field by two children so that his permission could be obtained although the women were already waiting for their interviewers. After he appeared, however, he did not give permission by simply expressing his word of authority, but requested the interviewer and interpreter to ask the women if they were willing to answer the prepared questions. The women indicated their willingness and in the same breath dictated very self-confidently the conditions for the discussion--they wanted to give a group answer after discussing among themselves first. The chairman sat with us during the discussion (which lasted from 9:15 a.m. to 11:15 a.m.) without trying to influence the women in any way (see table 4.6).

These were young, extremely lively women, and they conducted the discussion impulsively, yet discussed all the points thoroughly, with many participants weighing each. As the women had indicated, they then gave a group answer. Despite her youth and though she had far fewer children than the other women in the group--otherwise a sign of low prestige --Maimuna was the person most broadly recognized as the

TABLE 4.6

PARTICIPANTS IN THE WOMEN'S DISCUSSION GROUP

| Women | Age | School Grade Completed | No. of Children |
|---|---|---|---|
| 1. Rehema | 20 | 4 | 3 |
| 2. Hawa | 24 | 2 | 6 |
| 3. Zaina H. | 32 | 2 | 5 |
| 4. Zaina I. | 35 | - | 5 |
| 5. Sabia | 28 | - | 5 |
| 6. Maimuna | 20 | 6 | 2 |
| 7. Takia | 22 | - | 4 |

discussion leader. Her suggestions were usually accepted, perhaps because she had more schooling. Zaina H. was the most intensive participant in the discussion, but was somewhat controversial. Her attempts to dominate were not accepted by the others. In her rather quiet way, Takia was the discussion's third authority.

The answers were precise and came after brief and vehement counsel. The question about family planning in particular was answered so assertively that it embarrassed the chairman. This question was singled out for serious, objective considerations without the mild agitation that is often observed in discussion with men when a tabooed subject is touched upon.

To summarize the results of the two discussions, the women felt that people should have fewer children; that is, only as many as they could feed since otherwise they suffered with their starving infants. Even so, the women did not intend to plan their own families for the sole reason that this was against the wishes of their husbands. (The chairman became uncomfortable here while the women expressed their opinions frankly and freely as best as they could.) If they were to decide for themselves and if their husbands were to agree, they would not have any more than six children. Maimuna even said she would have only four. They could easily feed and clothe this number. The health of the mother was not acknowledged as a reason for limiting the size of the family since families that can feed and clothe themselves adequately would also have healthy mothers.

Shortly after 11:00 a.m. the participants became restless and moved to end the discussion so that they could prepare the noon meal. They did, however, bring up on their own a previous offer to be photographed for T.Shs. 10/-. When one considers that photography is otherwise disapproved of, this reaction is another reflection of "plain common sense."

Occupational needs. In dealing with the choice of occupation and the motivation for this choice, the following questions assumed that there were no special limitations such as education. The hypthothesis was that a large part of the rural population living in poverty would gladly improve its fate by practicing a well-paid and prestigious occupation if that were possible. On the other hand, one of the native experts interviewed in the study predicted that the responses would concentrate on the two occupations at the lower end of the scale--farmer and laborer--depending on the interviewee's level of education. The second expert presumed that four of the occupations would be considered worth striving for. That of laborer was one of them, because of the monthly income. That of businessman and merchant was another, for both could afford a pleasant life and possess the prestige symbol of a truck. The occupation of teacher would also be favored because of the relatively pleasant activity with a regular salary. The fourth occupation, that of a politician, would be desirable not only because of its high income but also because of the respect that is associated with it. The occupational priorities of 45 interviewees in 12 discussion groups were established (questions 03-07, see appendix and table 4.7).

TABLE 4.7

HIERARCHY OF OCCUPATIONAL DESIRES

| Occupation | Hypothetical Ranking | Ranking by Experts | Ranking by 45 Interviewees |
|---|---|---|---|
| Politician | 1 | + | 1 |
| Teacher | 2 | + | 3 |
| Businessman | 3 | + | 2 |
| Tailor (artisan) | 4 | no response | no response |
| Worker | 5 | ++ | 4 |
| Farmer | 6 | + | 1 |
| Other | 7 | no response | 4 |

+ positive

++ very positive

Unexpectedly, an absolute majority of the interviewees opted to keep their occupation as farmers. As the reason, the overwhelming majority cited independence and the security of providing one's own food. "One is free and does not have to follow any orders like a wage earner or a teacher. In general, everyone ultimately depends on agriculture. Even after retiring, you return to farming." Other, partially overlapping, reasons were tradition, well-being, self-employment, and the freedom to make decisions oneself. In all, more value was attached to the aspect of providing for oneself than of earning money: "You cannot eat clothing." "You can eat even before you sell."

The position of businessman, or actually of middleman, was desirable because of its high chances for profit and its aspect of self-employment, i.e., independence. Asked why they did not practice their dream occupation, the interviewees answered that lack of capital prevented it. The joy of teaching other people was regarded as the primary motive for choosing teaching as an occupation. They did not pursue that profession because they did not have the qualifications. There was only one occupational choice outside the given alternatives: cattle raising. It produces milk for home consumption and for the market as well as dung, meat, and skins. On the other hand, cows seem to have lost their significance as a dowry; paying cash is preferred. The fourth occupation--the politician's--was considered desirable because the individual in that field was in the position to represent the interest(s) of other people.

The interviewees were not always able to deal with hypothetical construction and thus to pretend that they had the necessary qualifications for the desired occupation. Nevertheless, the reasons cited indicated that the agricultural occupation was seen in a positive light and was consciously chosen, primarily for its aspects of subsistence and self-support. Those who know the country point out that this has not always been the case. Just ten years before, village children merely shook their heads in embarrassment when asked by their teacher what their fathers did. Since Tanazania's independence, the importance of agriculture and respect for the farmer's occupation has constantly been emphasized in information campaigns, adult education, and in school instruction. As a cross-check, a series of motives were presented in the following questions (see table 4.8). In the judgment of experts, the primary "material" needs were emphasized (food, income); social needs were attributed no importance. What was important is that striving for recognition and admiration--if it exists--was not admitted.

The twelve discussion groups cited those motives for their imaginary (or actual) choice of occupation that they felt were relevant for their positive decision. The motives esteemed most were those that could be associated with the need for income and employment. In addition, there was the consideration of the family, to be attributed to the need for belongingness and love. These four motives were chosen by eleven of the twelve groups. Amazingly, previous data appear to indicate that subsistence income was not preferred to cash income. The reason may be that this question did not ask about priorities but about positive influences on

TABLE 4.8

HIERARCHY OF MOTIVATIONS FOR CHOICE OF OCCUPATION

| Occupational Motivation | Hypothetical Ranking | Ranking by Experts | Ranking by 45 Interviewees |
|---|---|---|---|
| More food | 1 | ++ | 1 |
| More money | 2 | ++ | 1 |
| Work for the government | 3 | + | 10 |
| Regular income | 4 | ++ | 1 |
| Savings | 5 | + | 9 |
| (Old) age pension | 6 | no response | 6 |
| Group work | 7 | no response | 6 |
| Admiration and prestige | 8 | no response | 10 |
| Finding happiness in one's work | 9 | no response | 5 |
| Family cohesiveness | 10 | + | 1 |
| Village community | 11 | no response | 8 |

\+ positive ++ very positive

decisions in general.

The additional unstructured response item (question O8) was a check on how the earned money was used. Extra food was mentioned nine times, clothing eight times. Household utensils were cited three times, and, in the women's groups, spices and jewelry were mentioned twice each. The importance of providing food was reflected clearly in the answers to this question, although it concerned the use of money only. The interviewees felt that the monetary domain should not only supplement but also improve the satisfaction of subsistence needs.

The pragmatic list drawn up by both of the women's groups interviewed beautifully detailed what is meant by vital needs. For food, one buys cornmeal (*ugali*), rice, potatoes, bananas, beans, fish, meat, as well as spinach, papaya, oranges, eggs, milk, tomatoes, and onions. For cooking and for the house, one also needs soap, kerosine, oil or cooking fat, and sugar. Clothing is needed for the children, but also for the women themselves. This includes shoes, kanga and kitenge (the colorful, traditional dress of East African women), European-style clothing, underwear, including brassieres, not to mention jewelry like rings, bracelets, earrings, or necklaces. Clothing or gifts for the husbands were not listed. The very question elicited great commotion and violent disgust. The men received their own income from the coffee fields and should therefore be the ones to bestow gifts. Besides, who was to know if your husband might some day leave you and offer the gifts to some other woman?

The other motives which were to some extent significant for the choice of an occupation were enjoyment of the work, affiliation to groups and the village, and an old-age pension. These matters pertain to the operationalized needs of self-actualization, belonginess and love, and social security. It is surprising that positively valued old-age pensions are associated with coffee plantations. The reason is simply that a coffee field represents a long-term investment from which one can live in old age by leasing it or hiring workers. Bank savings, activity for the good of the country, and striving for admiration, were all outright unattractive as motives for choosing a particular occupation. But this result does not seem surprising. One does not think about savings until one's basic needs are satisfied. The good of the country or of the government is, realistically, subordinated to one's private needs, while the desire for admiration either has no function in the choice of an agricultural occupation or is, as already mentioned, tabooed to a certain degree.

Leisure needs. The experts who were interviewed presumed that drinking and conversation were the favored leisure activities. In fact, conversation with friends was by far the most frequent answer on the unstructured response item number 10. For reasons already cited, drinking was not mentioned at all. In comparison to the central activity, all other leisure needs were peripheral: people liked to listen to the radio, smoke, relax, do handiwork, or read the newspaper or books as far as this was possible. While sharing the interest in handiwork, the women added doing their hair and darning clothes. Further details are contained in the answers to the subsequent restricted response questions on leisure behavior (question 11).

The top position was accorded relaxation together with friends or family--in other words, satisfaction of belongingness and love needs. There followed the need for self-actualization by getting a better education. As with the other responses, it is naturally difficult to separate the desire from the actual circumstances. The results

TABLE 4.9

HIERARCHY OF LEISURE NEEDS

| Activity | Hypothetical Ranking | Ranking by Experts | Ranking by 45 Interviewees |
|---|---|---|---|
| Drinking pombe (beer) | 1 | ++ | no response |
| Talking with friends | 2 | ++ | 1 |
| Staying with the family | 3 | no response | 3 |
| Working in the field | 4 | no response | 6 |
| Earning money | 5 | + | 5 |
| Doing honorary work | 6 | + | 4 |
| Adult education or vocational training | 7 | ++ | 2 |

+ positive   ++ very positive

should therefore be interpreted as if the actual "use of one's leisure time" did not completely contradict perceived desires and needs, as if two factors were more or less in harmony with each other. This is shown well in the case of adult education. In the Shashui study area, there was a relatively thorough system of courses which were held for two hours, three times a week. What is more, the importance of reading and writing was repeatedly stressed by those attending. Among the honorary activities performed for the party or the government were group village tasks including Ujamaa farming and construction. One interviewee mentioned his military training. Here, too, obligation and voluntarism are intertwined. Yet while communal activities were not absolutely compulsory, their political importance was quite often emphasized. Thus the need for respect and prestige could be satisfied to a certain extent by participating.

Occupational activities were chosen just as frequently. This is, once again, an indication that the dichotomy between occupation and leisure is primarily a product of the industrial society and cannot necessarily be transferred. As will be shown, however, there were very definite ideas about the amount of labor time necessary for agricultural activity. The point at which the leisure activities must be cut back in favor of the need for material security and employment does have a bearing in this matter.

<u>Need priorities between labor and leisure time</u>. There are three issues to interrelate. First, the duration of work is to be determined. This means that the responses must be checked by means of observations and interviews with experts. Second, the farmers involved must be asked if they think that a higher labor input in the fields would increase crop yield in the first place. If such a result is not considered possible--regardless of whether the opinion is justified or unjustified--then related recommendations are useless. Finally, there is the question of subjective valuation. Which is the higher good, the direct satisfaction of needs in the leisure activity or the indirect satisfaction of needs motivated by producing more with additional labor? Of course, the question of values can logically be explored only if the possibility for increasing production is acknowledged. It, too, is a subjective parameter of strategies for raising productivity. Thus this

section makes the transition from the need orientation to the analysis of conditions that will determine an increase in agricultural production under given circumstances.

Figures given by native experts for the daily work time of the farming population ranged from five to nine hours. The information obtained from the interviewees fell precisely within this range, too. There was broad agreement that these statements applied for six days of the week, the Moslem Friday and the Christian Sunday not being included in the work week. Nevertheless, none of these figures seems any too conclusive as long as the qualitative grounds for determination are missing. The data were based in large part on work in the field and ignored other necessary tasks, especially the housework performed by the women. Similarly, seasonal fluctuations were largely ignored and must also be checked. Lost working hours were usually attributed to illness, death, or rainfall, if not omitted altogether. Thus, the trips to market were apparently included in the working time again or, because they occurred regularly, were not considered. More reliable here seems to be the paradigmatic description of a work day, the course of which was explored in detail during some of the discussions and was checked against the experience of an American research team (Fleuret and Fleuret) living in a village and against the statements of a missionary who had been residing in the area for many years (Br. Mathias, Sakarani Mission).

According to the observation and judgment of experts, most Shambaa are quite diligent people who usually work six to eight hours a day without a lunch break in the hot season under strenuous conditions. Women work somewhat fewer hours in the fields, but they are also responsible for keeping house (washing, cleaning, cooking, taking care of the children), and, above all, gathering firewood each day (Fleuret and Fleuret, 1976). Heavy loads are carried exclusively by women and girls. The wife arises early, at about 6:00 a.m., lets the livestock out of the house, and has already cleaned the stall by the time the husband appears around 7:00 a.m. She prepares a breakfast of *uji* (mush) with tea, takes care of the children, washes the laundry, and cleans the house. Around 9:00 a.m. the farmer goes into the field, the wife following an hour later with the children. All children who do not go to school help in the field. Even the three-year-old has a duty--looking after the one-year-old. If the little one cries, the mother comes to breastfeed and interrupts her work to quiet the baby, returning to hoe after half an hour. At midday, they all simply snack on something. Between 4:00 p.m. and 5:00 p.m., the wife and children return home, but not without carrying tremendous amounts of firewood. The very small children carry the hoes. When the father comes home at dusk around 7:00 p.m.--without carrying anything--the main meal of the day has already been cooked: *ugali* (corn mush with side dishes of vegetables or meat). After supper, the family sits together and tells stories. Projects like weaving mats or baskets are also worked on. Bedtime is about 9:30 p.m.

Although the daily routine apparently does not leave much extra time, eight of the twelve groups were of the opinion that it was possible to work more (question 12; see appendix). Three groups disagreed, and no answer was given on one occasion. Still, the positive result registered by these responses must be relativized with the reasons given. One group doubted the correlation between harder work and higher yield altogether. Two groups wanted to increase work by increasing communal activity or by enlisting more neighborly assistance. Two other groups gave no reason. Only three groups spelled out their positive responses in detail: One of these thought it possible to increase yield and income by working harder (more intensely) while the other two thought longer working hours (Group 3 specified 1 1/2 extra hours) would achieve the desired goal.

To interpret these answers, the results of other studies on the expenditure of labor in African regions (Brandt 1974) must be examined. According to these investigations, a full-time farm worker in Africa spends 900 to 1,100 hours in agricultural labor a year--apart from seasonal fluctuations and rainfall. In the most favorable situation, represented by a dry period of only two months, work time is distributed over 250 four-hour days. With a dry period of seven months, worktime is distributed over 150 seven-hour days. The apparent invariability of the upper limit on the average expenditure of labor is striking. It seems to indicate that more work is either physically impossible or economically pointless--as was assumed by half of the groups interviewed in the present study. The explanations suggested by Brandt go in both directions and also take into consideration the division of labor between husband and wife and how much of the "free" time, the time not used for working, is demanded by social activities:

- climate, illness, undernourishment and malnutrition
- alternative occupation like trading, cottage industry, gathering, hunting, fishing
- sex-specific division of labor
- social activities

Table 4.10 illustrates the relationships by presenting the breakdown of the time available to the farming population of a village in southern Uganda from February 1968 to January 1969.

The survey and observation in Shashui also verified the rigidity of the upper limits on labor, but this limit seemed to be much higher than 1,000 hours a year. To be precise, it was closer to 1,750 hours a year spent on agricultural work (250 seven-hour days). This difference is plausible because the physical strain caused by climate or illness is significantly less in the Usambara mountains than the corresponding conditions in the moist plains. This assumption is confirmed by Clayton (p. 440, cited in Brandt, 1974, p. 105), who in 1970 reported that 1,800 hours were worked per year in the Kenyan highlands, which have a similar climate. According to Brandt, only 500-700 hours a year are required for the full-time farm worker to achieve the subsistence minimum (Brandt, 1974 p. 104).

The difference between the work actually performed and the work necessary for subsistence needs allows hypotheses to be formulated about when and how much one can

TABLE 4.10

AMOUNT OF DAY (IN PERCENTAGES) SPENT IN SOUTHERN UGANDA PERFORMING SPECIFIC ACTIVITIES

| Activity | Men 30-50 Years | Women 20-50 Years |
|---|---|---|
| Agricultural work | 23.0 | 25.0 |
| Nonagricultural occupation | 6.0 | 0 |
| Housework | 2.5 | 19.0 |
| Eating, drinking, resting | 38.0 | 36.0 |
| Social activities | 30.5 | 20.0 |
| Total | 100.0 | 100.0 |

SOURCE: H. Brandt, "Zur Steigerung der Beschäftigung in afrikanischen Landgebieten--das Arbeitsangebot," _Zeitschrift für ausländische Landwirtschaft_ 13 (1974): 97-110, p. 101.

reckon that farmers will be prepared to work more. "The closer the production value is to the subsistence minimum, the more strongly the individual labor input is likely to react to incentives" (Brandt, 1974, p. 104). Vice versa, this means that the expenditure of labor will no longer react positively to increases in income near the upper limit of work input, which is determined by the favorable conditions in parts of the Usambaras. The tendency will more likely be to take the income exceeding subsistence needs and invest it in the cultivation of cash crops with higher work productivity.

From this emerge two possible lines of development for the labor supply, both of which result in making labor, alongside land and capital, a scarce production factor, which may sound like a paradox for a country with a seemingly inexhaustible reservoir of labor. In the first case, unfavorable exchange relationships between agricultural labor and industrial consumer goods raise the preference for leisure, that is, the need for spare time, and thereby limit the supply of agricultural labor. In the other case--and this might well apply to a far greater extent to the study area of Shashui--worsening environmental conditions caused by soil erosion and an absence of rainfall increase the expenditure of labor necessary to satisfy subsistence needs near the physical maximum.

The responses to the subsequent item (question 12.2) as to what was more important--having more free time or working harder--verified the results upon which the needs priority scale is based. All groups (one nonresponse) opted for the harder work. Leisure time had no value as an alternative to work in a cultural setting or in an economic situation where one can eat only what has been planted and nurtured by one's own hands. The reasons given point to this issue: "Harder work means a larger harvest and, thus, more to eat." "One must work hard to improve living conditions for next year." "You wouldn't harvest anything if you opted for leisure time." And: "It is better to work hard in order to achieve something than to stay idle and get nothing." These comments do not touch on the very highly ranked social needs observed and manifested in the questionnaire on "leisure time" needs as a conversation with friends and as family spirit.

## Comparing the Results of the Interviews with Objective Indicators and Data on Material Requirements

If the results from the above questionnaire were to be relevant for planning, they needed to be checked against empirically established "objective" indicators, data, and survey results.

The survey had been preceeded by a few years of drought, so that *food* and *cash reserves* in the villages had dwindled to a minimum. In addition to ongoing agricultural extension services, the district was even conducting food aid programs in extremely dry areas.

The *clothing situation* of the men and children was wretched while traditional costumes gave the impression that the women were well-dressed even on workdays. Clothing was in particularly short supply during cold periods in July and August, when the temperatures at the higher altitudes drop to the freezing point. Here there were almost no warm clothes like pullovers, jackets, or coats. Festival clothing for traditional holidays was usually available, however.

In terms of *shelter*, certainly no slum-like conditions were found anywhere. By local standards the average of four rooms for one family of mother and children was usually adequate. Almost half of the houses in the investigation area had a corrugated iron roof, though only one in fifteen had a cement floor (see the overall survey for the situation in Shashui in chapter 3). On the other hand, there was a lack of these features, particularly with the dwelling for each of the father's other wives. The traditional structures made of branches, leaves, and mud were far less durable, lasting about three years, and offered no effective protection from the heavy rain in the wet seasons.

In addition to the unhealthy living conditions that prevail in some seasons, the sanitation facilities in the villages were rudimentary, as only a few villages had their own water supply and none had a sewage system. On the other hand, the *medical care* was good. For emergencies, Lushoto and Bambuli both had a hospital within walking distance. There were many dispensaries, although some were poorly supplied with medicines.

The *educational system* was even better than the medical service. The capacity of the primary schools was great enough to admit every child, if not at six years of age, then at eight or nine. The state primary schools were supplemented by Koran schools and other educational opportunities in Christian missions and parishes. For adults there were courses in reading and writing, schooling for mothers, and other advanced education policy plans to provide elementary education to as much of the population as possible. On the other hand, the possibilities for educating an elite beyond this level were severely circumscribed by a limited curriculum.

The typical family possessed one hectare of *arable land*, which was extremely little, particularly because one could not then afford to regenerate a part of the soil by leaving it fallow. Four-fifths of the population lived exclusively from agriculture while every tenth person worked as a hired laborer and one out of fifteen was self-employed (chapter III). Income and the circulation of money in Shashui were correspondingly limited.

Most *transportation* was by foot. This was true primarily for buying and selling at the market places as well as for the daily trips to the field or to school. Leisure was taken at dusk. At that time, friends and family sat together and chatted. Journeys to distant markets and visits to friends in far-off villages can be called diversionary "trips." There are signs that such trips have become less frequent over the last two generations. The necessary work time has increased as land has become scarcer. In addition to state and religious holidays, the death of a family member or a friend played a large role in social life. The entire hamlet, and often neighboring settlements as well, took part in the ceremonies, some of which lasted for days.

*Environmental hazards* do not have the peripheral significance often ascribed to them in industrial countries; rather, these are central to the farmer's very existence. The most important problems are drought, a drop in the level of ground water, soil erosion, and the degeneration of soil and vegetation into semi-desert.

## Evaluation and Conclusions

A few theoretical limitations on the validity and utility of the results must be acknowledged a priori when evaluating the status quo of a needs situation--in the present case, that of a sector or the population of the Usambara Mountains in the year 1976. Within this limitation, though, the needs concept can provide valuable suggestions for participative planning. But to substitute needs research for a strategy of participation would be to overestimate the potential of the former. It is astounding how inaccurate some of the estimates of the native experts were--an argument in favor of participative planning.

First, the concept that satisfaction of a need calls forth the next "highest" need and that needs are thus hierarchically ordered, appears extremely questionable. Certainly there is nothing to be said against the heuristic assumption that need potential is in principle infinite. But one cannot assume a quasi-mechanical sequence since the manifestation of each need is influenced by all other needs, including the individual and social situation of the interviewee. Lujo Brentano recognized this as early as 1908 when he wrote that a human being requires the satisfaction of his physical needs in a way that corresponds to his simultaneous spiritual, ethical, aesthetic, mental, and social expectations (Brentano, 1908, p. 58). In such a limited sense, the positions taken on every need ranking can and should be evaluated only in a manner that stresses acute urgencies under the assumption of a broad *ceteris paribus* domain: Because of misery and scarcities, the physiological needs for food, clothing, and shelter clearly take top priority.

Another point to consider is the cultural and sociohistoric circumstances not only of the needs but of the research concept based on them. Within western industrialized countries themselves, for example, the need for social interaction ranks much higher in Japan than in the United States or Western Europe. With esteem and self-actualization it is just the opposite (Schoess, 1977). This presents no problem as long as cultural prejudices do not invade the research concept and veil the culture-specific differences. The current theory based on Maslow was developed for Western industrialized countries, where the satisfaction of needs is integrated into the production process in such a way that labor is seen as a means to satisfy needs largely, if not exclusively, during leisure time. "Needs are historical products dependent on the structure of society and the state of society's development" (Brede and Siebel, 1977, p. 10). In other words, needs in industrialized countries are confined to leisure time. "If things went only according to needs research, people would only be consumers"(Brede and Siebel, 1977, p. 15). In order to avoid this sociohistorical prejudice in the present study, work time needs and leisure time needs, paradoxically, had to be separated precisely because such a dichotomy might not exist. Only in this way can they be discussed.

Keeping the aforementioned limitations in mind, the following needs structure can be discerned:

Social relationships are not created in such a way that they can admit of the difference between alienated work and consumer-oriented free time. The subsistence economy prevailing in the investigation area permits neither dependency on wages nor cash income for any significant level of consumption. Nevertheless, there is a clear division between work time and work-free time in the minds of the population, though this division is a result of needs that are different from those of industrial countries. Farming in the fields is a response to--and in large measure a satisfaction of--all basic needs listed above. Some of the physiological needs, those for food and shelter, are met without an exchange process. The safety need is met by the highly valued factor of independence. Social contact is fostered by group work just as the village community provides for esteem and prestige, but satisfaction of these needs within the work process seems to play a subordinate role in comparison with the other needs already cited. In spite of the hardships related to the climate and soil, finding enjoyment in work has a relatively high value, and it is a sign for self-acutalization in a self-directed activity.

As expected, the somewhat less frequently noted "higher" needs, primarily the social need for love and belongingness, were satisfied during leisure hours in contact with family and friends. In contrast, the psychic needs for self-actualization in the form of further training and prestige in the form of taking on honorary work proved to be much weaker.

The responses concur with the demand to be noted externally as well, but provide essential additional details. At this point, the subject of work motivation leads into both the following chapters--the agricultural production process on peasant farms and the resulting choice of production strategies and their possibilities for boosting output.

## CHAPTER V

# AGRICULTURE ON PEASANT FARMS AS A BASIS FOR A STRATEGY OF "SELF-RELIANCE"

## Crops

A wide variety of cash and subsistence crops is cultivated in the Shashui study area, the most important being described below (see also Tanga Plan, vol. 2, pp. 82-102). The distinction between "cash" and "subsistence" refers only to the primary purpose of cultivation. Naturally, produce intended for domestic consumption is sold at the local market, too, when there is a surplus. The most important cash crops are vegetables, coffee, and tea. The most important subsistence crops are corn, beans, bananas, sweet potatoes, and cassava.

### Cash Crops

Vegetables. The Usambara mountain range is the most productive vegetable growing area in Tanzania. In 1973, 8,600 tons were harvested, of which a large part were marketed in the cities of Dar es Salaam and Tanga. The most important types were white cabbage and tomatoes, followed by potatoes, carrots, onions, cauliflower, and cucumbers. A few of these have found their way into the native cuisine. Potatoes, for example, are highly prized as a staple; cabbage is occasionally boiled; tomatoes and onions spice the meat broths. As important as vegetables are as a source of vitamins for the native population, they are expensive for the farmers who do not grow their own. The potato is a good example to use for profitability calculations. At 35 cents per kilogram, gross profits on them are about T.Shs. 720/- per hectare or T.Shs. 11/- per workday (see table 5.1).

Coffee. Arabica coffee is grown in many parts of the Usambara mountains. The round, sweet berries are picked when red, indicating that they are ripe. Subsequently, the seed is extracted, fermented, washed, and sun-dried on large gratings or mats.

The small farmers grow coffee extensively without spending too much time and effort in plowing, weeding, mulching, or applying pesticides. Coffee berry disease (CBD) can kill up to 20 to 30 percent of the crop. The main risk, however, is posed by storms and climate. Insufficient rainfall destroys the crop.

Bananas are customarily planted between the coffee plants. Some agricultural experts claim that this reduces the distance between the plants too much, that the bananas produce excessive shade and rob the soil of important nutrients. Ecologists, on the other hand, argue that the coffee also benefits from the shade and that it tolerates the bananas very well, as, for example, on Mount Kilamanjaro. In any case, the normal coffee harvest of 100 kilograms per hectare in the Usambaras is extremely small.

A different problem is the low remuneration the farmers receive. In the years preceding this investigation, the cooperatives probably withheld the final payments which were due after sales. Since 1976, the National Milling Corporation

TABLE 5.1

GROSS PROFITS PER HECTARE/ACRE PER WORKDAY FOR VARIOUS CROPS (IN T.SHS.)

| Enterprise | Production Method | | | | | | | | |
|---|---|---|---|---|---|---|---|---|---|
| | Traditional | | | Stage I | | | Stage II | | |
| | Gross margin/ hectare | Gross margin/ acre | Gross margin/ workday | Gross margin/ hectare | Gross margin/ acre | Gross margin/ workday | Gross margin/ hectare | Gross margin/ acre | Gross margin/ workday |
| Coffee | 264/- | 108/- | 5/74 | 445/- | 182/- | 5/16 | 780/- | 320/- | 8/01 |
| Tea | 744/- | 305/- | 6/88 | 850/- | 348/- | 4/89 | 1,880/- | 774/- | 7/25 |
| Irish potatoes | 721/- | 296/- | 10/92 | 992/- | 407/- | 8/55 | 682/- | 280/- | 6/82 |
| Sweet potatoes | 500/- | 205/- | 6/94 | 750/- | 307/- | 5/56 | 1,025/- | 420/- | 6/14 |
| Beans | | | | | | | | | |
| - irrigated | 865/- | 355/- | 10/30 | 1,059/- | 434/- | 8/83 | 1,415/- | 580/- | 9/25 |
| - nonirrigated | 490/- | 200/- | 7/- | 701/- | 287/- | 7/08 | 1,057/- | 433/- | 8/59 |
| Maize | | | | | | | | | |
| - local & improved | 482/- | 198/- | 4/82 | 794/- | 325/- | 5/36 | 797/- | 327/- | 5/73 |
| - hybrid | - | - | - | 480/- | 197/- | 3/78 | 930/- | 381/- | 6/29 |
| Cassava | 1,170/- | 480/- | 12/72 | 1,480/- | 607/- | 11/21 | 1,665/- | 683/- | 11/32 |
| Bananas | 675/- | 277/- | 12/98 | 1,240/- | 508/- | 11/27 | - | - | - |

SOURCE: The United Republic of Tanzania, Regional Development Office, Tanga Regional Development Plan 1975-1980, 3 vols. (Tanga: Tanga Integrated Rural Development Programme, 1975), 3: 258. (Mimeographed.)

NOTE: Errors in the original have been corrected by the author.

(MNC) has transported the coffee from the villages, and the cooperatives have been dissolved.

Under the given production conditions and with a normal selling price of T.Shs. 3/- per kilogram (by 1973), gross profits per hectare of cultivated land amount to T.Shs. 5/70 per workday (see table 5.1). The authors of the Tanga Plan estimated that intensification of coffee growing with the help of monoculture cultivation and greater pest control would increase soil productivity greatly to T.Shs. 780/-, but would boost labor productivity only slightly to 8/-. The recommendation to build an electric pulping factory in Soni (Tanga Plan, vol. 3, p. 217) should be considered with caution, however. A manual factory like the one in Shashui would cost less and would be less prone to mechanical failures.

Tea. The importance tea growing has for peasants is still increasing, particularly in view of the fact that the second five year plan made it a national priority. The Tanzania Tea Authority has increased the amount of land devoted to tea in the highlands by 1,600 hectares since 1963. A tea factory was built in Mponde to process the harvest. New areas like Mtimule, Mzungu, and Mangika in the Shashui study area were also opened for growing tea.

Nevertheless, 85 percent of the tea is still produced on 75 percent of the land available to the tea estates, where the average yield is 1,500 kilograms per hectare. In spite of strict supervision by the Tanzania Tea Authority and its network of advisors--a staff of 36 in 1973--the peasants do not always observe the procedures required for intensive cultivation. These include fertilizer use and weeding in particular. The output is commensurately low--250 to 300 kilograms per hectare. Even so, there is a gross yield of T.Shs. 750/- per hectare or T.Shs. 7/- per workday (table 5.1). The Tanga Plan estimated that 600 kilograms per hectare (T.Shs. 1,800/- per hectare or T.Shs. 7/25 per workday could be achieved with better cultivation and intensive use of fertilizers. Although the peasants do not understand the necessity of intensifying labor for non-food crops, the net profit of tea is still attractive. Tea growing is much more lucrative than coffee growing as long as drought does not destroy the crop altogether.

## Subsistence Crops

Corn. Corn (maize) is the key crop. As a staple food, it fetches a good price on the local markets. It was not until the 1920s that corn surpassed sorghum as the primary food. Traditionally, this place in the diet was occupied until the last century by the cooking banana, which, unlike sorghum, is still eaten today. The area in Lushoto devoted to corn production is estimated to be 10,000 hectares.

Farmers plant either native seeds or the high-yielding hybrid corn, which they buy. Land preparation and planting are both time-consuming activities. Each kernel is planted separately and is fertilized with dried cow dung if any is available. Traditionally, peasants attempt to reduce the risks involved by placing five kernels in each hole--probably because germination of the local seeds is less reliable. The farmers rationalize this wasteful procedure by saying that they are caring for their fellow creatures: A kernel for the rat, a kernel for the bird, and, lastly, one for man. Consequently, several corn stalks often grow together, compete for nutrients, and ultimately produce only small ears of corn. Yet it would be outrageous to every farmer to thin out plants which are already growing, even if this action would increase the yield.

The output of 650 kilograms per hectare could be tripled through mechanization. However, at a price of T.Shs.-/75 per kilogram, the gross yield of T.Shs. 480/- per hectare would not even be doubled because of the high costs such an approach would incur (see table 5.1).

Beans. Beans serve as supplementary food and can help to remedy the widespread protein deficiency. They also fetch a high price on the market. Another advantage is their brief maturation period. Lastly, they lend themselves extremely well to intercropping, being combined mostly with corn, since as legumes, they enrich the soil with nitrogen.

When traditional methods of production are used, the yield of beans reaches 330 kilograms per hectare. With irrigation it rises to 550 kilograms per hectare, which, at a prescribed sales price of T.Shs. 1/75 per kilogram makes beans very competitive with other produce (table 5.1).

Bananas. Bananas are not grown in a monoculture; they are usually planted together with coffee. The banana portion of the crop can occasionally be even more important to the farmer since it caters to his individual consumption needs. Cooking bananas are dried and processed into flour from which a paste called *ugali* is made. Very little work time is needed to cultivate bananas, which makes it possible to achieve a high degree of labor productivity.

The main steps in the work are planting and harvesting. Pest are negligible. The trees are trimmed occasionally. The yield per hectare is reported to be 250 to 350 bunches (4-6 tons), and, according to estimates of the Tanga Plan, it is supposed to reach 650 to 950 bunches (10 to 15 tons) in a monoculture. Because of the small expenditure of work, labor productivity is T.Shs. 13/- per workday. With effective fertilizer use, mulching, irrigation, and a favorable distance between the trees, monoculture should bring in T.Shs. 1,200/- per hectare or T.Shs. 11/- (!) per workday (table 5.1).

Sweet potatoes. Although sweet potatoes are grown primarily for home consumption, they are often sold as well. The total area devoted to sweet potatoes in Lushoto is not known. A yield of 1.5 tons produces a profit of T.Shs. 500/- per hectare. An increase up to the modest yield of 3 tons per hectare would mean a gross profit of T.Shs. 750/-. Labor productivity would, however, then sink from T.Shs. 6/90 to T.Shs. 5/50 (table 5.1).

Cassava. Cassava plays a less important role in the Usambara mountains than in the plains. Its significance lies in its suitability for storage. Cassava is simply left in the ground until it is needed. Thus, there was always a food reserve (famine crop), which led to the practice--and not just in colonial times--of always planting a certain amount of cassava.

Cultivating cassava requires little effort. Thorough preparation of the soil is not necessary and weeding is rarely needed. It is not simply the crop's resistance to drought that makes it attractive; it also fetches a high market price. During the eighteen month period of growth, one hectare of land produces roots that may amount to only two metric tons of dried chips when processed, but which guarantee a gross profit of T.Shs. 520/- per hectare or T.Shs. 10/- per workday and more (table 5.1).

As with other edible roots, the biggest problem for cassava is the destruction caused by apes during the day and wild boar at night. Since keeping constant watch would be too labor-intensive, cassava is usually planted only within the village or in the immediate vicinity of the house where people are nearly always present. This problem has changed little since the days of Mbegha, the hunter and legendary first king of the Shambaa, 200 years ago.

By increasing labor (with types of cassava that are less susceptible to certain viruses) it should be possible to produce a yield of 4-10 tons, which would correspond to a gross profit of T.Shs. 1,500/- to 1,800/- per hectare and year or T.Shs. 11/- per working day. Enlarging cassava cultivation would then be worthwhile even in the Usambara mountains, in spite of the fact that climatic conditions are not as favorable for cassava there as on the plains.

## Peasant Farms

### Structure

Agricultural area. About 12 percent of the total area of the Lushoto district is used for agriculture. Of this, individual farms account for three-quarters of the agricultural land (see table 5.2).

TABLE 5.2

AGRICULTURAL LAND USE IN LUSHOTO, 1973 (IN 1,000 ha.)

| | |
|---|---|
| Estates | 6.3 |
| Individual farms | 31.6 |
| Ujamaa villages | 4.3 |
| Arable land | 42.2 |
| Total area | 345.2 |
| Arable land as percent of total area | 12.2% |

SOURCE: The United Republic of Tanzania, Regional Development Office, Tanga Regional Development Plan 1975-1980, 3 vols. (Tanga: Tanga Integrated Rural Development Programme, 1975), 2:58. (Mimeographed.)

Farm size. Of the 31,600 hectares of agricultural area belonging to individual farms, 90 percent is utilized; in other words, this land is very intensively used. More than half of the area is accounted for by tiny farms of less than half a hectare. Only 0.5 percent of all individual farms have more than five hectares of arable land (see chapter 3 and table 5.3).

TABLE 5.3

INDIVIDUAL FARM SIZES IN LUSHOTO, 1973 (IN PERCENTAGES)

| Size | Percentage of Farms in Lushoto |
|---|---|
| Less than 0.5 hectares | 54.0 |
| 0.5 - 0.99 | 19.4 |
| 1.0 - 1.99 | 16.8 |
| 2.0 - 2.99 | 7.4 |
| 3.0 - 3.99 | 1.9 |
| 4.0 - 4.99 | 0.3 |
| 5.0 or more | 0.5 |
| Total number of farms (rounded off) | 100.0 |

SOURCE: The United Republic of Tanzania, Regional Development Office, *Tanga Regional Development Plan* 1975-1980, 3 vols. (Tanga: Tanga Integrated Rural Development Programme, 1975), 2:60. (Mimeographed.)

*Labor force*. In 1973 there were 90,700 people who were either employed as family laborers on individual farms or self-employed. Thus, each family had an average of 2.17 workers, or one worker for every 0.37 hectares of cultivated land.

Because of this situation and because of the high annual rate of population growth (3.2 percent), underemployment in the countryside and the number of landless people are both expected to rise. A more effective use of land and labor with the aid of more efficient agricultural technologies thus seems urgent in the short run. In spite of such efforts, estimates assume that the carrying capacity of the Lushoto district will be reached in the 1980s (*Tanga Plan*,

1975-1980, vol. 2, p. 67).

Carrying capacity. The carrying capacity serves as an index for the relative over- or underpopulation of an area, whereby a certain level of technology is presupposed (Moore, 1971). If subsistence farming with traditional methods is assumed, the carrying capacity is a direct function of the soil's productive potential, including the necessary proportion of fallow land. With regard to cash crops, price differences must be taken into account regionally, over time, and in relation to other crops. The Tanga Plan's calculations assumed an energy need of 2,000 calories per person per day and a gross cash income of T.Shs. 2,500/- per five-member family per year. Neither differential production costs nor technological changes were included (see table 5.4).

TABLE 5.4

POPULATION CAPACITY OF THE LAND RELATED TO ACTUAL POPULATION

| Area | Estimated Population Capacity | Ratio: Population capacity/ Actual Population | |
|---|---|---|---|
| | | 1967 | 1975 (est.) |
| Lushoto District | 139,000 | 0.66 | 0.52 |
| Tanga Region | 728,000 | 1.02 | 0.84 |

SOURCE: The United Republic of Tanzania, Regional Development Office, Tanga Regional Development Plan 1975-1980, 3 vols. (Tanga: Tanga Integrated Rural Development Programme, 1975), 2:334. (Mimeographed.)

The situation has worsened throughout Tanzania, but it is the mountain regions that have suffered most from acute land shortage. According to the assumptions above, the Usambara mountains are overpopulated, although this does not mean that the people there are starving. The horizontal mobility and social security mechanisms like farming in the plains, emigration, use of land at no cost, and family and clan ties prevent that. Still, minimal income was set quite high, substantially higher than the cash income actually earned from agriculture. At any rate, the inevitable conclusion is that agricultural productivity must be boosted and that nonagricultural jobs must be created.

## Factors of Production

Land. As already shown, the general scarcity of land means for the average independent farmer that the availability of arable soil is barely sufficient to feed his family.

Labor. The work is done primarily by family members. The father works seven to nine hours a day in the field, six days a week, sometimes even going without a break at noon. Moslems do not work on Fridays or Christians on Sundays.

Women and the older children also work in the fields, but normally only until noon. The afternoon is reserved for doing work in and around the house. Smaller children drive what cows the family may have into the meadow. The women are usually responsible for subsistence farming while the men take care of the cash crops. Heavy labor like tillage is always done by the men, however.

The work peaks that occur with tilling, weeding, or harvesting are handled by various forms of neighborly help. *Ngemo* refers to unilateral aid that does not involve cash. A group of men work together for a neighbor; in return, the latter serves them an evening meal and, if they are not Moslems, the alcoholic beverage, *pombe*. Then there is *kuimyana* for mutual aid, whereby only two or three friends help each other. Serving food is not necessary, but reciprocal assistance is obligatory.

Wage labor is rare in the countryside but it does exist. It is considered a first step toward the creation of a rural petty capitalist or kulak class and is thus frowned upon by the government. The minimum wage, set by the state, for a worker in the countryside, is T.Shs. 230/- per month in 1976. Payment is often determined by performance, however.

Official group work is normally carried out for two days a week by members of the Ujamaa villages in the communal field, where cash crops--like tea in the Ujamaa village of Mtimule--are usually cultivated. Subsistence cultivation is reserved for the family farms, though fields of hybrid corn worked by cooperatives are occasionally found. Contrary to principle, Ujamaa members are also compensated separately for special tasks like cattle keeping and nonagricultural skilled tasks.

Capital and investments. The supply of capital is minimal. Obtaining it even for buying seed, fertilizer, and insecticides is a problem. No credit is provided to the peasant. However, the Ministry of Agriculture does support the cultivation of coffee and tea by making available seedlings, NPK fertilizer, fungicides, and insecticides whose costs are later deducted from the profits. Ujamaa villages receive these inputs free of charge.

There are no investments for mechanization, transportation, and storage. Farming is still done with the traditional way with the *jembe* (hoe) and the *panga* (machete). Middlemen or--as long as it existed--the rural development project LIDEP picked up the vegetables with trucks. Coffee was either taken to the cooperative in Soni or collected in the villages by the National Milling Corporation (NMC). All other produce, including surplus vegetables, was usually carried on the heads of women. It was taken on certain days to be sold at familiar market places, as far away as Mombo and Mazinde at the foot of the Usambara mountains (A. Fleuret, 1976). Small dealers transported their vegetables or fruit by using the bus that runs to Tanga or Dar es Salaam.

Technological progress Neither plow, team of oxen, nor tractor were used in the Usambaras. In any case, they would be inapprorpiate for the steep slopes and scattered fields that are encountered in most of this area. Other possibilities for mechanization are locally developed handtools that simplify and accelerate work. Their use would promise greater productivity. Further advances in productivity can be achieved with the help of an improvement in farming methods.

## Selection of Farms

Criteria. Theoretically, the following critera are considered when selecting the types of farm to be studied:

- altitude
- farm size
- degree of cooperation
- technological level
- crops
- animal husbandry

Certain combinations of criteria would then serve as parameters for ideal types of farm units. Examples might be:

- a large irrigated farm that uses agrochemical aids and monocultural farming;
- a small farm that has no irrigation or chemical input and prefers intercropping.

Selected farms. The general household survey (chapter 3) indicated, however, that the socioeconomic structure of the investigation area of Shashui was relatively homogeneous. There were indeed differences in property and income, but these were not so large that typical large farms would have had organizational structures and farming methods fundamentally different from typical small farms. Within a range of more or less uniform techniques, the pattern of methods varied only slightly, depending on capital input. If no money was available, then less insecticide was used and spraying intervals were extended and vice versa. The pattern of methods were determined much more by the crops which were being cultivated. Vegetables required a different approach to that used in coffee, corn, or tea production. The strategy was adapted to climatic variations associated with the diverse altitudes in the area.

Consequently, the farms and individual fields included here were selected primarily according to primary and secondary crops. In addition, the five subareas within the study area, which represent the entire spectrum of climatic zones and crops in the Usambara mountains, are considered. The following analysis describes and evaluates the inputs and outputs of the selected farms.

## Costs and Returns

The purpose of this section is--after having established existing needs and available resources--to examine the farms themselves in order to discover, on the basis of agricultural production, if there were still economic reserves to be utilized.

If the material requirements in the subsistence area and the market situation in the sales area are taken as given, then the economic reserves affect the farms' inputs and outputs. The inputs correspond to the use of means of production and the resulting costs, whereas the outputs are the yields or, measured in given market prices, the returns. Some problems were that economic evaluations of domestic farms did not yet exist for Lushoto and that the observation period was too brief to allow reliable surveys of our own. Hence, the data were gathered by observation (measuring, counting) as far as was possible, but data on the yields were supplemented by interviews.

Before continuing, however, the problems associated with a technology-dependent increase in yield, especially

FIG. 5.1. HIERARCHY OF AGRICULTURAL COSTS

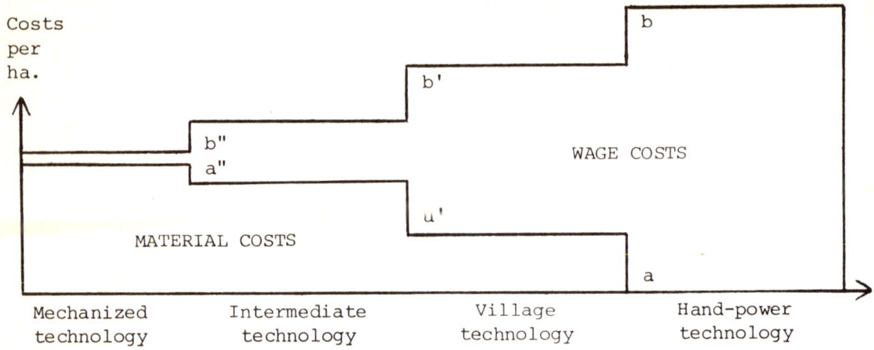


SOURCE: After G. A. MacPherson and D. Jackson, "Village Technology for Rural Development--Agricultural Innovation in Tanzania," International Labour Review 111 (1975): 97-118, p. 115.

based on marginal economic variations applied can be considered, but these variations would no longer be measured quantitatively in monetary units. The benefits would be represented in a qualitative scale of preferences instead (duration of working time, leisure time, desire for a greater variety of food).

One example of greatly diverging combinations of inputs is a potato field in Limuru (Kenya) where a variety of technologies was used. A corresponding comparison of costs was published by the GAT in Nairobi in a guide booklet for potatoe growing (Growing Potatoes in Kenya, 1974, pp. 18-19). The labor time required by the recommended procedures for growing potatoes is calculated and presented in table 5.5. For plowing one acre and planting the potatoes, for example, the farmer needs 5 male workers for 6 days or 3 male workers for 10 days, in short, 30 workdays. Each spray requires 2 workdays; land preparation and weeding, 10; and harvesting, 25.[1]

---

[1] G. Elwert and D. Wong (1978), pp. 96-109) follow a similar approach in calculating production conditions.

TABLE 5.5

AMOUNT OF LABOR NEEDED PER ACRE PER SEASON FOR GROWING POTATOES

| No. of weeks | Operation | Workdays |
|---|---|---|
| 1 | Field preparation (hoe cultivation) | 15 |
| 2 | Planting | 15 |
| 3 | Spraying | 2 |
| 4 | Ridging/weeding 2X spraying | 14 |
| 5 | 2X spraying | 4 |
| 6 | Ridging/weeding 2X spraying | 14 |
| 7 | Weeding 2X spraying | 14 |
| 8 | 2X spraying | 4 |
| 9 | Spraying | 2 |
| 10 | Spraying | 2 |
| 11 | Spraying | 2 |
| 12 | Spraying | 2 |
| 13 | Harvesting | 25 |
| | Total workdays needed per season | 115 |

SOURCE: German Agricultural Team in Kenya, Growing Potatoes in Kenya, edited by the Federal Agency for Economic Cooperation [B.f.E.] Division Agriculture (Essen: Webels, 1974), p. 18.

If one assumes that the work procedures (field preparation, planting, and harvesting) occur under the same conditions, then the question arises as to how land preparation and weeding relate to the work time needed for spraying pesticides and applying fertilizer and if they can be partially substituted for one another. Thirty workdays per acre per season are set aside for spraying and 30 days for plowing. These cannot be substituted and, hence, technological alternatives are not obvious from reading the table. At best, they must be deduced from the subsequent cost-benefit comparison

The following model for comparing methods, taken from the brochure cited above, includes three ways of planting, and compares their estimated costs and profits per acre (see table 5.6). The preconditions below are used in the model's calculation.

- The potato sacks are procured for T.Shs. 3/- and are used used for six harvests; they are not sold with the potatoes.
- Salaried workers make up 50 percent of the labor force and are paid T.Shs. 3/- per workday.
- The price of potatoes is set at T.Shs. 20/- per sack.

According to the calculations of the model, the farmer with the greatest profit (III) is the one who puts the most money into cultivation, the one whose operation is the most capital intensive. He uses all the prescribed amounts of inputs like chemical fertilizers, pest control products, and high yield seed. Thus he pursues the capital intensive agricultural technology required by the "green revolution" policy. Farmer II, by contrast, uses domestic seed that is less expensive but that has significantly less potential. What is not taken into consideration is that this seed is quite possibly more resistant to pests and perhaps would need only a reduced spraying schedule. Farmer I comes off worst. He relies solely on locally available resources. He is the only one to register a loss, although the spraying periods are eliminated and thus need not be included in a calculation of costs.

TABLE 5.6

CALCULATION MODEL FOR COSTS AND GROSS RETURNS PER ACRE OF POTATOES PER SEASON (IN T.SHS.)

| | Farmer I | Farmer II | Farmer III |
|---|---|---|---|
| Yield, bags/acres (90 kg each) | 18 | 90 | 135 |
| Gross returns | 360/- | 1,800/- | 2,700/- |
| Local seeds (10 bags/acre) | 250/- | 250/- | - |
| Certified seed (10 bags/acre) | - | - | 600/- |
| Fungicides | - | 130/- | 130/- |
| Pesticides | - | 70/- | 70/- |
| Fertilizer | - | 200/- | 200/- |
| Transport (1/50 per bag) | 27/- | 135/- | 202/- |
| Costs of bags (-/50 per bag) | 9/- | 45/- | 67/- |
| Total variable costs | 286/- | 830/- | 1,269/- |
| Gross margin before costs for casual labor deducted | 74/- | 970/- | 1,431/- |
| Casual labor costs | 120/- | 165/- | 165/- |
| Net margin after costs for casual labor deducted | 46/- | 805/- | 1,266/- |

SOURCE: German Agricultural Team in Kenya, Growing Potatoes in Kenya, edited by the Federal Agency for Economic Cooperation [B.f.E.], Division Agriculture (Essen: Webels, 1974) , p. 19.

Along with the calculation in monetary terms, one would assume that the population's food supply is guaranteed best by a larger harvest. But on this matter a few points must be raised. The price of potatoes can fluctuate, usually downwards, when the supply increases. According to the present calculation model, there is the stimulus for producing more. As usual, the entrepreneurial risk increases with the opportunities and is thus at its greatest when capital input is maximal. Risk would, however, be possible to bear under certain circumstances if uncertainties were confined only to profits, which depend on the economic formation of price and on the ecological and climatic situation (manifest in the size of harvest).

But the above-mentioned brochure was written before the first oil crisis, and since then the greatest uncertainties have arisen on the *cost* side. Prices for oil products have soared. Prices for fertilizer in developing countries, about a third of which was subsidized by the Tanzanian government, tripled within a year. Ultimately, wages do not stagnate, either, in a period of rising prices and expectations.

The model presented by GAT describes a stable and favorable combination of all input and output factors; it is a "best-case" model. To estimate the magnitude of entrepreneurial risk--not the probability of its occurrence--a corresponding "worst-case" model would have to be calculated on the same basis.

One "worst-case" model (variation A) begins by making a few realistic assumptions. There is a moderate inflationary development in the urban industrial sector. The general labor cost increase of 20 percent, including social benefits, is passed on, resulting in a general price increase of 10 percent. The stagnating agricultural sector, particularly the subsistence sector, whose prices for agricultural products and seed remain constant, is an exception. On the other hand, the family farm operating within the subsistence economy has no labor costs. The price of imported products (hybrid seed) increases by 50 percent because of the unstable currency exchange rates. Furthermore, imported chemical products increase by an average of 200 percent because of the global increase in the price of crude oil. Table 5.7 presents the figures calculated on the basis of this increase in costs.

In spite of a seriously aggravated cost situation in the relation of the various agricultural techniques to each other, the profit ratios of the worst-case, *variation A*, remain more or less intact. The unrealistic assumption that labor costs money was avoided only for the small farm operated in the traditional manner.

In *variation B*, the cost structure of variation A is preserved, but at the same time the yield and profit situation worsens slightly, again not an unusual situation. Slightly less precipitation reduces the yield by 10 percent. Simultaneously, influences of the world market bring about a modest 10 percent price adjustment downwards. This alters the calculation as shown in table 5.8

Because of the less favorable cost and profit situation, all three agricultural or farm types find themselves in the margin where cultivation is barely or no longer profitable. Of course, it is striking that the losses in this variation increase disproportionately with increasing capital intensity--just as the profits did before. Capital-intensive farm III is the only one operating in the red.

TABLE 5.7

WORST-CASE MODEL VARIATION A: COSTS AND GROSS RETURNS OF POTATOES IN AN UNFAVORABLE COST SITUATION (IN T.SHS.)

| | Farmer I | Farmer II | Farmer III |
|---|---|---|---|
| Yield, bags/acre (90 kg. each) | 18 | 90 | 135 |
| Gross returns | 360/- | 1,800/- | 2,700/- |
| Local seed (10 bags/acre) | 250/- | 250/- | - |
| Certified seed | - | - | 900/- |
| Fungicides | - | 260/- | 260/- |
| Pesticides | - | 140/- | 140/- |
| Fertilizer | - | 400/- | 400/- |
| Transport (1/50 per bag) | 30/- | 148/50 | 222/- |
| Costs of bags (-/50 per bag) | 10/- | 49/50 | 74/- |
| Total variable costs | 290/- | 1,248/- | 1,996/- |
| Gross margin before costs for casual labor are deducted | 70/- | 552/- | 704/- |
| Casual labor costs | - | 198/- | 198/- |
| Net margin after costs for casual labor are deducted | 70/- | 354/- | 506/- |

SOURCE: Based on the German Agricultural Team in Kenya, Growing Potatoes in Kenya, edited by the Federal Agency for Economic Cooperation [B.f.E.], Division Agriculture (Essen: Webels, 1974).

TABLE 5.8

WORST-CASE MODEL VARIATION B: COSTS AND GROSS RETURNS PER ACRE OF POTATOES IN AN UNFAVORABLE COST AND YIELD SITUATION (IN T.SHS.)

| | Farmer I | Farmer II | Farmer III |
|---|---|---|---|
| Yield (bags/acre = -10%) | 16.2 | 81 | 121.5 |
| Gross returns (-10% or 18/- per bag) | 291/60 | 1,458/- | 2,187/- |
| Material costs and transport (total variable costs) | 290/- | 1,248/- | 1,996/- |
| Gross margin | 1/60 | 210/- | 191/- |
| Casual labor costs | - | 198/- | 198/- |
| Net profit/loss | +1/60 | +12/- | -7/- |

SOURCE: Based on the German Agricultural Team in Kenya, Growing Potatoes in Kenya, edited by the Federal Agency for Economic Cooperation [B.f.E.], Division Agriculture (Essen: Webels, 1974).

*Variation C* of the worst-case model makes two additional assumptions. The price of artificial fertilizer has not doubled, but instead, has tripled--i.e., it has been passed on in full, not subsidized by the state. Moreover, it is assumed that farm III, operating along modern lines, is a large farm that must pay all its hired workers in cash. With no change in the other assumptions of variation B, there arise cash wage costs totalling T.Shs. 414/- on the basis of 115 workdays per acre and a daily wage of T.Shs. 3/60. A cost-returns comparison between a large farm operating on a capital and wage intensive basis, a medium-sized farm run more along modern lines, and a small traditional family farm is presented in table 5.9.

TABLE 5.9

WORST-CASE MODEL VARIATION C: COSTS AND GROSS RETURNS PER ACRE OF POTATOES ON SMALL, MEDIUM-SIZED, AND LARGE SCALE FARMS WITHOUT STATE SUBSIDIES (IN T.SHS.)

| | Farmer I | Farmer II | Farmer III |
|---|---|---|---|
| Yield | 16.2 | 81 | 121.5 |
| Gross returns | 291/60 | 1,458/- | 2,187/- |
| Material costs and transport (total variable costs) | 290/- | 1,448/- | 2,196/- |
| Gross margin | 1/60 | 10/- | -9/- |
| Casual labor costs | - | 198/- | 414/- |
| Net profit/loss | +1/60 | -188/- | -423/- |

SOURCE: Based on the German Agricultural Team in Kenya, Growing Potatoes in Kenya, edited by the Federal Agency for Economic Cooperation [B.f.E.], Division Agriculture (Essen: Webels, 1974).

With uncertain expectations, the large farm operated on a wage and capital intensive basis carries a much larger risk of economic loss than the traditionally run small family farm. It must be remembered that the per acre profits or losses shown in the model calculation must still be multiplied by the size of the farm.

The premise in all variations was, as in the original example on which they are based, the application of a monetary perspective. From this it follows, first, that the monetary and subsistence case in the empirical investigation are comparable only under certain conditions. Second, it means that an all too hasty modernization of the subsistence economy involves great dangers for the country's supply lines. The advantage of the traditional family farm is that it does not have to earn the quasi fixed costs of paid labor and debt service--usually capital made available by means of outside financing--in order to avoid debts. If necessary, it can live off its own "equity" of family labor without effective interest.

Furthermore, the model in all its variations was developed with a short-term microeconomic point of view; the long-term ecological component, which leads to public support of the individual economic costs when symptoms of overcropping appear, have not been included at all. But even from the microeconomic viewpoint, the farm types I, II, and III are not based on genuine cultivation alternatives. They are not, in other words, based on qualitatively different production methods. The difference was only quantitative. From a complete, self-contained agricultural technology--the "green revolution" technology used by farm type III--successive inputs were simply deleted from the "alternative" or, better, "reduced variations" without substituting any others. Given the reality of the economic and ecological situation, the question arises as to the socioeconomic feasibility of genuinely different technologies in the broadest sense.

## Typical Farms--Two Case Studies

The following section introduces two complete agricultural farms in the Shashui study area whose locations and production programs were comparable but which were different in size and thus provided a different level of well-being for their "owners." They represent a typical affluent and less affluent Shambaa family. On the basis of the individual data from these and other plots of land, a comparison of the inputs and outputs will be made. From this one can derive ideal types of crop budgets and, according to the combination of crops, farm budgets as well. These data serve in the evaluation of existing production strategies and in chapter 6 as one of several bases for recommendations for the improvement of production strategies.

A wealthy farmer. Saidi lived in Ntuii and earned between T.Shs. 380/- (minimum wage) and T.Shs. 600/- a month as a night watchman in Soni. With this regular cash income he was automatically among the wealthier and more privileged farmers. For "once a farmer (*mkulima*), always a farmer" regardless of which sideline is pursued. As a Moslem, he was a member of the prevailing confession.

These few basic data almost automatically determined Saidi's socioeconomic situation. He was able to afford four wives, for each of whom he not only had to pay a high price at marriage but whom he also had to provide with fields and houses. With the help of the wives and children, however, he was able to cultivate considerably more land, which, in turn, increased his income. Saidi was 58 years old and fed eleven children. He had attended primary school, which means that he was at least able to read and write. With six rooms, his home is relatively spacious. Like the three houses in which his wives lived, it, too, was covered with a corrugated iron roof.

His four fields were larger than average, and they were not scattered too far from each other. Field number 1, which he worked most intensively, had coffee, bananas, yams (the only ones in the vicinity), and, recently, castor trees. It began directly in front of his door. The cow dung around the house was simply thrown into the field, where it automatically distributed itself throughout the soil. Abundant fertilizer was a by-product of animal husbandry, and its disposal, like that of kitchen wastes, demanded no great amount of time. As a good farmer, Saidi also took care to protect the soil by covering it with dry weeds. The mulch layer was supposed to decompose gradually and replenish the soil's fertility. Saidi did not need to buy artificial fertilizer. He used 1.5 liters of spray concentrate a year to control pests.

None of the three other fields were more than ten minutes away by foot. Situated on a hillside were fields 2 and 3, which grew the staple foods of corn and cooking bananas for homo consumption. Saidi was able to afford to have these fields hoed by paid laborers before field preparation, but the planting and harvesting was done by the family. Weeding and mulching was done by the father alone.

Field 4 was located on the moist valley floor. In this fertile location the profitable market specialty of the Usambara mountains (vegetables) was planted. Here, Saidi opted to grow the most popular and saleable crops--white cabbage and tomatoes.

The three crops in field 1--coffee, bananas, and yams --had to be considered as being equally important. Neither bananas nor yams could be called a secondary crop. The extra labor after growing coffee was very small (see table 5.10). For this reason, the gross margins per workday were actually too high. As is always the case with mixed cropping, the gross margin per workday as well as the area yield and the resulting gross margin per hectare must be seen integrally, that is, per field unit. The gross margin shown in table 5.10 represents subsistence income valued at market prices.[1] On the other hand, coffee yielded real cash income, though only a modest amount because of drought. But since capital input--as with bananas and yams--was practically nil, this reduction in yield had no disastrous consequences. This meant protection against risk. The fact that the peasants followed this risk-avoiding production strategy in the cash domain as well justified the researcher's approach of combinding subsistence income and cash income in table 5.10. Gross margins are almost identical with gross returns, though only under the condition that the price of the saplings is not included in this one-year calculation. With 110 plants at T.Shs. -/20 and lifespans of 20 years, the proportional costs would be only about T.Shs. 1/- per year. The labor input was reported as being astoundingly low for coffee, though, too. As described above, both weeding and mulching were performed at the same time. In addition, there was the distributing of diluted spray and the harvesting. The trees

---

[1]Measuring home consumption, too, in market prices is not a completely reliable undertaking. For the sake of comparison, however, this aspect should not be neglected, as peasants do regularly market their surplus subsistence crops locally.

TABLE 5.10

FARM BUDGET IN NTUII (FIELD 1)

| Field 1 = 2,401 $m^2$ | Crops | | | Total |
|---|---|---|---|---|
| | Bananas | Yams | Coffee | |
| | Long Rainy Season | Short Rainy Season | | |
| Yield | 50 bunches | 90 roots | 28 kg | |
| Price per unit | T.Shs. 10/- | T.Shs. 10/- | T.Shs. 5/- (1975) | |
| Gross returns | T.Shs. 500/- | T.Shs. 900/-[a] | T.Shs. 140/- | |
| Material input | | | | |
| Seeds | Farmer's own | Farmer's own | - | |
| Fertilizer | - | - | Regularly | |
| Manure | Together with coffee | Together with coffee | - | |
| Pesticides | - | - | T.Shs. 5/- | |
| Gross margin | T.Shs. 500/- | T.Shs. 900/- | T.Shs. 135/- | T.Shs. 1,535/- |
| Labor input (in days) | | | | |
| Field preparation (hoeing) | - | - | - | |
| Planting | - | 4.5 | - | |
| Weeding | Together with coffee | Together with coffee | 4 | |
| Mulching | Together with coffee | Together with coffee | Simultaneously with weeding | |
| Fertilizing | Together with coffee | Together with coffee | - | |
| Spraying | - | - | 1 | |
| Pruning | 6.5 | - | - | |
| Harvesting | 1 | 25 | 4 | |
| Workdays | 7.5 | 29.5 | 7 | 44 workdays |
| Gross margin per workday | T.Shs. 66/67 | T.Shs. 30/51 | T.Shs. 19/29 | T.Shs. 34/89 |

[a] Market price in the event of a sale

were not trimmed in the first five years. Fertilizing, which is so important for the yield, occurs incidentally and was not calculated as labor input. Moreover, the ten weeks needed for preparing the land, digging holes, and planting would in a calculation of averages have to be spread over 20 years, which would total three days per year.

The family--as many as four wives and their children--assisted in working the land in both corn fields (see tables 5.11 and 5.12). In field 3, ten to twenty neighbors occasionally helped, too. That is the reason for the varying levels of gross margins per workday, which are no longer comparable, especially with the result achieved by corn in vegetable field 4 (see table 5.13). On the other hand, it would have been futile to try and ascertain the exact total amount of labor expended. What remains is a calculation at the subsistence level for a family unit. The help received from the neighbors is thus offset, as the workdays gained on the family's fields are lost when returning the service for the neighbors. In keeping with this procedure, the case study likewise included the paid laborers in subsistence terms (i.e., workdays) for accounting purposes. In this case, too, one wife and two children helped with harvesting vegetables.

The total gross margins for all four fields without figuring the labor input was T.Shs. 2,752/50. The gross income of the peasant Saidi thus approached the upper income limit of T.Shs. 3,000/- estimated for the Usambara mountains. Except for coffee and vegetables, this was essentially subsistence income or income in kind. Because of the low level

TABLE 5.11

FARM BUDGET IN NTUII (FIELD 2)

| Field 2 = 3,680 $m^2$ | Crops | | Total |
|---|---|---|---|
| | Bananas | Corn | |
| | | Short Rainy Season | |
| Yield | None; planting took place in preceding year | 375 kg | |
| Price per unit | - | T.Shs. 1/- | |
| Gross returns | - | T.Shs. 375/-[a] | |
| Material input (in T.Shs.) | | | |
| Seeds | Farmer's own | T.Shs. 35/- | |
| Fertilizer | - | - | |
| Manure | - | - | |
| Pesticides | - | - | |
| Gross margin | - | T.Shs. 340/- | T.Shs. 340/- |
| Labor input (in days) | | | |
| Field preparation (hoeing) | - | 25[b] | |
| Planting | - | 2 | |
| Weeding | - | - | |
| Mulching | 3 | 12 | |
| Fertilizing | - | - | |
| Spraying | - | - | |
| Pruning | - | - | |
| Harvesting | - | 3 | |
| Workdays | 3 | 42 | 45 workdays |
| Gross margin per workday | - | T.Shs. 8/10 | T.Shs. 7/56 |

[a] Market price in the event of a sale

[b] One worker for T.Shs. 200/-

TABLE 5.12

FARM BUDGET IN NTUII (FIELD 3)

| Field 3 = 3,174 $m^2$ | Crops | | Total |
|---|---|---|---|
| | Bananas | Corn | |
| | | Short Rainy Season | |
| Yield | 36 bunches | 245 Kg | |
| Price per unit | T.Shs. 10/- | T.Shs. 1/- | |
| Gross returns | T.Shs. 360/-[a] | T.Shs. 245/-[a] | |
| Material input (in T.Shs.) | | | |
| Seeds | - | - | |
| Fertilizer | - | T.Shs. 30/- | |
| Manure | - | - | |
| Pesticides | - | - | |
| Gross margin | T.Shs. 360/- | T.Shs. 215/- | T.Shs. 575/- |
| Labor input (in days) | | | |
| Field preparation (hoeing) | - | 12[b] | |
| Planting | - | 2 | |
| Weeding | - | 1.5 | |
| Mulching | - | - | |
| Fertilizing | - | - | |
| Spraying | - | - | |
| Pruning | 7.5 | - | |
| Harvesting | 0.5 | 2 | |
| Workdays | 8 | 17.5 | 25.5 workdays |
| Gross margin per workday | T.Shs. 45/- | T.Shs. 12/29 | T.Shs. 22/55 |

[a] Market price in the event of a sale

[b] One worker for T.Shs. 160/-

TABLE 5.13

FARM BUDGET IN NTUII (FIELD 4)

| Field 4 = 1,428 $m^2$ | Crops | | | Total |
|---|---|---|---|---|
| | White Cabbage | Tomatoes | Corn | |
| | Short Rainy Season | Short Rainy Season | Long Rainy Season | |
| Yield | 8 bags | 125 kg | 70 kg | |
| Price per unit | T.Shs. 15/- | T.Shs. 1/20 | T.Shs. 1/- | |
| Gross returns | T.Shs. 120/- | T.Shs. 150/- | T.Shs. 70/-[a] | |
| Material input | | | | |
| Seeds | T.Shs. 15/- | T.Shs. 7/50 | T.Shs. 15/- | |
| Fertilizer | - | - | - | |
| Manure (peasant's animals) | - | - | - | |
| Pesticides | - | - | - | |
| Gross margin | T.Shs. 105/- | T.Shs. 142/50 | T.Shs. 55/- | T.Shs. 302/50 |
| Labor input (in days) | | | | |
| Field preparation (hoeing) | 6 | Included | 12 | |
| Planting | 25 | Included | 1 | |
| Weeding | 9 | Included | 6 | |
| Mulching | - | - | Simultaneously with planting | |
| Fertilizing | - | | Simultaneously with field preparation | |
| Spraying | - | - | - | |
| Pruning | - | - | - | |
| Harvesting | 1 | 1.5 | 2 | |
| Workdays | 41 | 1.5 | 21 | 63.5 workdays |
| Gross margin per workday | T.Shs. 5/82 | | T.Shs. 2/62 | T.Shs. 4/76 |

[a] Market price in the event of a sale

of capital input, the cash part of this could be used as a kind of surplus for purchasing clothing or household utensils. In spite of its participation in the monetary economy, Saidi's manner of production tended to be a risk-minimizing subsistence strategy.

A poor farmer. Abrahamani from Ngalani pursued no vocation other than farming. He possessed only two adjacent fields with a total area of less than 2,100 square meters (see tables 5.14 and 5.15). He was 37 years old and had one wife and two children. Having had no primary school education, he learned to read and write as an adult. With four rooms and a thatch roof, his one house met traditional standards only. He possessed no cows or other domestic animals except for a few chickens. He used no fertilizers or pesticides.

His bisected field was located at the place where he lived. Abrahamani grew coffee on two-thirds of his land with some bananas in mixed cultivation. On the remaining one-third he grew corn during the long rainy season and tomatoes in the short rainy season.

Because of the drought, his coffee yield was very low. In 1976 the year of the study, Abrahamani expected a yield of 250-300 kilograms which was to bring him T.Shs. 1,600/-, for prices had risen since 1975, when the climatic circumstances depressed his gross earnings to a reported total of T.Shs. 400/-.

TABLE 5.14

FARM BUDGET IN NGALANI (FIELD 1)

| Field 1 = 1,350 $m^2$ | Crop: Coffee |
|---|---|
| Yield | 40 kg. |
| Price per unit | T.Shs. 5/- |
| Gross returns | T.Shs. 200/- |
| Material input | |
| Seed | - |
| Fertilizer | - |
| Manure | - |
| Pesticides | - |
| Gross profit | T.Shs. 200/- |
| Labor input (in days) | |
| Field preparation (hoeing) | - |
| Planting | - |
| Weeding | 9 |
| Mulching | - |
| Fertilizing | - |
| Spraying | - |
| Pruning | 6 |
| Harvesting | 9 |
| Workdays | 24 workdays |
| Gross margin/workday | T.Shs. 8/33 |

TABLE 5.15

FARM BUDGET IN NGALANI (FIELD 2)

| Field 2 = 736 $m^2$ | Crops | | Total |
|---|---|---|---|
| | Corn | Tomatoes | |
| | Long Rainy Season | Short Rainy Season | |
| Yield | 70 kg | 108 kg | |
| Price per unit | T.Shs. 1/- | T.Shs. 1/20 | |
| Gross returns | T.Shs. 70/- | T.Shs. 130/- | |
| Material input | | | |
| Seeds | Farmer's own | Farmer's own | |
| Fertilizer | - | - | |
| Manure | Amount unknown | Amount unknown | |
| Pesticides | - | - | |
| Gross margin | T.Shs. 70/- | T.Shs. 130/- | T.Shs. 200/- |
| Labor input (in days) | | | |
| Field preparation (hoeing) | 38 | 4 | |
| Planting | 2 | 3 | |
| Weeding | 12 | 4 | |
| Mulching | Simultaneously with weeding | Simultaneously with weeding | |
| Fertilizing | 3 | 3 | |
| Spraying | - | - | |
| Pruning | - | - | |
| Harvesting | 6 | 15 | |
| Workdays | 61 | 29 | 90 workdays |
| Gross margin per workday | T.Shs. 1/15 | T.Shs. 4/48 | T.Shs. 2/22 |

## Crop Budgets

For the purpose of gathering microeconomic data relating to peasant subsistence farms in Shashui, a total of sixty fields were surveyed. Of these, eighteen were measured and their dimensions compared with the data obtained from the owners. Information on the inputs and the outputs was given by farmers from all five subareas, which included thirteen of the thirty-two hamlets in the study area. Information on the prices for inputs and on market prices for crops was easily verified. Information on amounts was checked against comparable statements and against information obtained in interviews with agricultural experts.

The following paragraphs will present a few typical crop budgets based on the available data. This is a difficult venture since both the inputs and the outputs vary greatly. Both depend on the combination of crops, which differs from one farm to the next. For this reason the following procedure seems suitable. All fields surveyed are categorized according to their main crops--tea, coffee, vegetables, corn, and other crops. The various secondary crops like bananas, beans, yams, cocoyams, and cassava are excluded from the initial calculation of inputs and outputs. The traditional gross yield as well as gross margins per hectare and workday are presented separately. Because of the variety of possibilities, no average values can be given here. Instead, an attempt is made to present paradigmatic examples of farming situations according to the groups of main crops. By "paradigmatic" is meant that, in spite of qualitative and quantitative variations, the situations cited in each case are the ones most likely to be found, either as described or in very similar form.

Tea. Tea had not been introduced as an ujamaa activity in the village of Mtimule until 1973-1974. However, in 1976 the situation was still not typical and was difficult to assess, for only 100 kilograms were reaped in the first harvest of 1975, and nothing at all was harvested in 1976 because of the great drought. Tea production in this area thus had to be considered an agricultural innovation. Cultivation on family farms was just beginning in 1976. The first attempts at this also took place in 1974, but the drought thwarted positive results for the first three years.

Investments in the communally worked field of several hectares ran to T.Shs. 14,300/- from 1974-1976. The largest part of this was devoted to the procurement of 26,000 tea plants. In addition there were seven sacks of fertilizers a year at T.Shs. 50/-. In the first year, 89 members of the tea commune spent an estimated 4,700 workdays preparing the field, planting, and weeding. The commune was able to reduce this input in the second year to 2,600 days. The year of the first harvest took only 750 workdays, and in the drought year of 1976, 280 workdays were devoted to tea cultivation. The projected yield per hectare fluctuates according to the intensity of farming between 1,000 and 4,000 kilograms of tea leaves. At a delivery price of T.Shs. -/56 per kilogram in 1975, which soared the following year, respectable gross returns reaching T.Shs. 2,200/- are where labor must be substituted by capital, will be illustrated. This will be done using a calculation model based on a recommendation from the German Agricultural Team (GAT) in Kenya.

## Model of Costs and Returns Dependent on Technology as an Input Factor

A hierarchy of economic perspectives is useful when considering rural technologies. This differentiates between mechanized, intermediate, and village technology, as well as hand power, with labor cost per hectare rising and the costs for capital falling as mechanization decreases. "The most advanced technologies require less total input but more cash input than do the less advanced ones" (MacPherson and Jackson, 1975, p. 114). For the farm embedded in the monetary economy, that is, one that hires workers and sells produce, the technological input varies according to changing costs. If with a high degree of mechanization the savings of labor costs (b) exceed the excess expenditure for capital service (a), then "more advanced" technology is favored (see figure 5.1).

Of decisive importance for economically underdeveloped areas--including much of East Africa--is the division of the technology which lies between mechanization/motorization and hand power into an intermediate and further simplified village technology that can be used within the village environment and with the village's own resources. Village technology will also be the form appropriate to the conditions of the subsistence economy. In this domain, the analysis

TABLE 5.16

CALCULATION FOR COFFEE AS THE PRIMARY CROP

| 1 Hectare of Coffee | Peasant Family Farm | Professionally Operated Estate |
|---|---|---|
| Yield | | |
| in the poor year of 1975 | 150-300 kg | 717 kg |
| in a year with adequate rainfall | 400-600 kg | 1,250 kg |
| Price per unit | T.Shs. 5/- | T.Shs. 10/- to 13/- |
| Gross returns | | |
| in the poor year of 1975 | T.Shs. 750/- to 1,500/- | T.Shs. 7,416/- |
| in a year with adequate rainfall | T.Shs. 2,000/- to 3,000/- | T.Shs. 12,355/- |
| Material input | | |
| Seeds | 400-700 saplings | 1,250 saplings |
| Fertilizer | - | T.Shs. 316/- |
| Manure | at times | - |
| Pesticides | T.Shs. 3/- | T.Shs. 453/- |
| Overhead costs | - | T.Shs. 648/- |
| Gross margin | | |
| in the poor year of 1975 | T.Shs. 747/- to 1,497/- | T.Shs. 5,997/- |
| in a year with adequate rainfall | T.Shs. 1,997/- to 2,997/- | T.Shs. 10,936/- |
| additional income from secondary crops grown with two methods of farming | T.Shs. 600/- to 1,000/- | - |
| Labor input (in days) | | |
| Field preparation (hoeing) | - | 27 |
| Weeding | 70 | 273 |
| Mulching | Simultaneously with weeding | - |
| ertilizing | - | 8 |
| praying | 5 | 22 |
| runing | 30 | 44 |
| arvesting | 10-30 | 32 |
| ocial benefits | - | 239 |
| Workdays | 115-135 | 645 = T.Shs. 6,646/- |
| Gross margin per workday | | |
| for coffee in 1975 | T.Shs. 9/- | T.Shs. 9/29 |
| for secondary crops in 1975 | T.Shs. 12/90 | - |
| Average | | |
| per hectare of coffee in 1975 | T.Shs. 12/25 | T.Shs. 9/29 |

possible under favorable weather conditions--and even more with higher prices.

In the Ujamaa village of Mtimule, a great deal of capital was infused and a strategy of profit maximization was pursued. The considerable risk involved could be carried only through the participation of the Tanzanian Tea Authority. The village inhabitants paid for the initial failures only with the loss of their labor input. In the course of time, this path will certainly produce development advantages for the village. The peasant tea producer, on the other hand, must proceed more cautiously. He suffers not only from the labor lost to his subsistence fields but also from the loss of the capital investment or the carry-over of debt incurred by the purchase of plants and fertilizer.

Coffee. Eight coffee fields ranging from 1,350 to 4,000 square meters were included in the study along with a professionally run plantation of about eight hectares (see table 5.16). The eight small farms were dispersed throughout the entire study area. Labor was provided by the family, which was occasionally supported by help from the neighbors.

Capital investment was limited essentially to the nonrecurring investment for the coffee saplings available at T.Shs. -/10. No commercial fertilizer was used. Manure was applied only to the extent that it was available, and whenever possible for planting the saplings. Rarely was more than one kilogram of spray concentrate used per hectare per year--diluted by a factor of 6. Aside from the expenditures on pest control, which were intially handled on credit and later charged against the harvest, there was no continuous capital investment.

The data on inputs/outputs were based on the year 1975, when weather conditions were unfavorable. In years with high precipitation--scarcely the rule anymore--the yield can be doubled or tripled. Field comparisons are difficult in either case since coffee is not grown in a monoculture. That is, a low coffee yield per hectare still does not necessarily mean a poor yield per plant and can be part of an overall satisfactory return per hectare.

The producer's price per kilogram for the mediocre quality in 1975 was T.Shs. 5/-. From 1973 to 1976 the price doubled from T.Shs. 3/- to T.Shs. 6/-. The heavy dependence on the world market, which can have short term negative effects as well, has a direct impact on the peasant farmers. For that reason it is a sign that a risk-balancing strategy is being pursued when additional income--usually subsistence income--is produced in the coffee field by growing corn, bananas, cassava, yams, or cocoyams on the same plot of land. Assessed at market prices, this additional income, distributed over two growing seasons, can account for a considerable proportion of the total income along with the cash income earned from coffee sales.

Under certain circumstances, more than half of the 120 to 180 workdays a farmer spends per hectare of coffee that is mixed with other crops can be devoted to the "secondary crops." On the whole the entire methods of farming seems to be excessively labor intensive. This is especially true for the high figures reported for pruning trees and picking berries. The time reported for picking was much higher in a few cases because the gradual process of ripening requires weekly harvesting. Data on field preparation and planting procedures are omitted here since only the one-year rhythm of an already mature field can and should be considered. Mulching and weeding were performed in one step.

The gross margin per workday was relatively favorable because of attractive coffee prices. At the same time, the low level of capital investment kept the risk small. Naturally, gross margin is much higher when there is sufficient rainfall, but then the labor required to work the field and to harvest is greater. Gross margins from the secondary crops per workday would in certain cases reach the immense sum of T.Shs. 30/- to 60/- since cassava, for example, needs little care and since no additional time is needed for weeding. Banana trees only need to be pruned. The gross margin per workday for the more labor-intensive cultivation of corn and beans was T.Shs. 8/-. The gross margins for additional workdays of T.Shs. 12/90, a figure obtained in the survey average for all secondary crops, raised the corresponding field margins of T.Shs. 9/- to T.Shs. 10/25.

The professionally run plantation was substantially more labor intensive and was more profitable in good years. In 1975, a poor year, it suffered a loss despite a gross margin of T.Shs. 6,000/- per hectare and 9/29 per workday. The reason lay in the high capital intesnity, which, because of the wage earners, included labor costs. Four-fifths of the total costs are attributable to wages that must be paid whatever happens. Family farms were run much more flexibly in this regard. If the work on a crop is no longer worth the effort, the family laborers can immediately be directed to a more profitable activity. Accordingly, the same is true for the high proportion of overhead costs incurred for transportation, correspondence, insurance, or activities like street improvements and night guard duty that are not directly productive. Vice versa, mulching is eliminated as part of the work procedure since there is not enough material available for it. Therefore, covering the soil, which is ensured on the small farm and which makes ecological sense, is not practiced.

Vegetables. Fourteen vegetable fields ranging from 120 to 3,200 square meters were studied, plus a so-called professionally cultivated field of 1,300 square meters (see tables 5.17 and 5.18). White cabbage and tomatoes were the most frequent crops, followed by lettuce, cucumber, and radish. The usual, though not the only, farming period for vegetables was the short rainy season. This means that the fields were tilled anew in August, when corn was harvested. White cabbage was harvested in November, tomatoes in January, so that the growing season lasted from four to six months. The peasant had to coordinate this with the market circumstances, a problem that even the extension service could not always help with satisfactorily. Very often there was a surplus of a particular crop, which severely depressed the price. During this study, such was the case with white cabbage, which had a very poor price, while tomato farmers were able to secure satisfactory prices. Since the peasant's cash income is ultimately

TABLE 5.17

CALCULATION FOR VEGETABLES AS THE PRIMARY CROP ON A PEASANT FAMILY FARM

| Field of 1 Hectare | Crops | | Total |
|---|---|---|---|
| | Tomatoes (Moneylove) | Cabbage | |
| | Short Rainy Season | Short Rainy Season | |
| Yield | 2,800 kg | 3,000 kg | |
| Price per unit | T.Shs. -/90 | T.Shs. -/30 | |
| Gross returns | T.Shs. 2,520/- | T.Shs. 900/- | |
| Material input | | | |
| Seeds | T.Shs. 100/- | T.Shs. 120/- | |
| Fertilizer | - | - | |
| Manure | When available | When available | |
| Pesticides | - | - | |
| Gross margin | T.Shs. 2,420/- | T.Shs. 780/- | T.Shs. 3,200/- |
| Labor input (in days) | | | |
| Field preparation (hoeing) | 60 | 60 | |
| Planting | 60 | 60 | |
| Weeding | 80 | 80 | |
| Mulching | - | - | |
| Fertilizing | Some use of manure during the planting process | Some use of manure during the planting process | |
| Spraying | - | - | |
| Pruning | - | - | |
| Harvesting | 55 | 45 | |
| Workdays | 255 | 250 | 505 workdays |
| Gross margin per workday | T.Shs. 9/49 | T.Shs. 3/12 | T.Shs. 6/34 |

TABLE 5.18

CALCULATION FOR VEGETABLES AS THE PRIMARY CROP ON A PROFESSIONALLY OPERATED FARM

| Field of 1 Hectare | Crops | | Total |
|---|---|---|---|
| | Tomatoes (money-maker) | Cabbage | |
| | Long Rainy Season | Short Rainy Season | |
| Yield | 5,000 kg | 13,000 kg | |
| Price per unit | T.Shs. -/90 | T.Shs. -/30 | |
| Gross returns | T.Shs. 45,000/- | T.Shs. 3,900/- | |
| Material input | | | |
| Seeds | T.Shs. 520/- | T.Shs. 240/- | |
| Fertilizer | - | - | |
| Manure | T.Shs. 25/- | T.Shs. 45/- | |
| Pesticides | | | |
| Killpest 50% | - | T.Shs. 504/- | |
| Dethane M 45 | T.Shs. 1,001/- | - | |
| Agral 90 (adhesive) | T.Shs. 187/- | T.Shs. 374/- | |
| Depreciation for implements | T.Shs. 612/- | T.Shs. 612/- | |
| Gross margin | T.Shs. 42,655/- | T.Shs. 2,125/- | T.Shs. 44,780/- |
| Labor input (in days) | | | |
| Field preparation (hoeing) | 112 | 112 | |
| Ridging | 114 | 93 | |
| Planting | 68 | 93 | |
| Weeding | 35 | 35 | |
| Cutting stakes | 68 | - | |
| Mulching | - | 70 | |
| Fertilizing | Manure applied in planting process | Manure applied in planting process | |
| Spraying | 57 | 23 | |
| Pruning | 23 | - | |
| Harvesting | 114 | 47 | |
| Workdays | 591 | 473 | 1,064 workdays |
| Gross margins per workday | T.Shs. 72/17 | T.Shs. 4/49 | T.Shs. 42/09 |

determined by the relationships between supply and demand, it seems pertinent to illustrate this dependence on market fluctuations with the example of the two typical crops, tomatoes and cabbage.

The yields on the peasant family farm were very small; the high prices for seed often prevented full utilization of the available land. In most cases the expenditure on seed was the only cash investment. Composted manure was used when available during the planting procedure. A second reason for the small yields was that lower yielding, cheaper types of seed were used, especially in the case of tomatoes. The professionally run farm used only high quality seeds, fertilized selectively with cow dung, and sprayed to control pests. The basic prices for vegetables were the delivery prices that the LIDEP development project paid to the farmers late in 1975. The prices for seed and pesticides were also cited from the LIDEP lists. Better quality and greater quantities of the inputs plus more intensive care increased the yield of professional farming immensely. One could count on 18,000 heads of cabbage at 700 grams each per hectare or 10,000 tomato plants, which could yield five kilograms per plant if the "money-maker" variety was used.

The professional farm was not only more capital intensive, but more labor intensive as well. Thus, driving support stakes or chemical pest control were added as extra work steps. Astonishingly, the small family farm eliminated the mulching that is traditional with other crops. The farmers had been told by the extension service to keep their fields "clean." In both types of fields, some of the work times reported seemed excessive, the reason being perhaps that the fields were small and far apart. Long trips and small errors in estimations have a strong effect when extrapolating to one hectare. Vice versa, the data reported for the professionally worked field excluded the work procedures that were not directly related to production. They would reduce the gross profits per workday of T.Shs. 42/- to less than 20/-.

In spite of the inevitable unreliability of the data and great variations, it is clear that the yields can be increased drastically with suitable technology. On the other hand, the factor ultimately responsible for the amount of income and yield per hectare or workday is the market price, which the individual farmer cannot influence. With a little care, the peasant farmer is able to achieve a satisfactory result so long as the delivery price does not fall to a very low level. Even the professional farmer cannot offset a fall in prices by drawing on the capital and techniques available. Gross margins of T.Shs. 4/- to 5/- per workday mean a loss of about T.Shs. 4,000/- per hectare since the laborers must be paid their full wage.

Corn. The study included ten fields, ranging in size from 575 to 4,000 square meters, on which corn was the primary crop (see table 5.19). The fields were situated primarily in subareas 1 and 2 (see chapter 3). Both submountainous areas (3 and 5), whose climate was unfavorable for growing corn, were excluded. Secondary crops were beans, bananas, cocoyams, and coffee.

TABLE 5.19

CALCULATION FOR CORN AS THE PRIMARY CROP ON A PEASANT FAMILY FARM

| Field of 1 Hectare | Long Rainy Season | | Short Rainy Season | Total |
|---|---|---|---|---|
| | Corn | Beans | Cocoyams | |
| Yield | 775 kg | 180 kg | 300 kg | |
| Price per unit | T.Shs. 1/- | T.Shs. -/75 | T.Shs. -/25 | |
| Gross returns | T.Shs. 775/- | T.Shs. 135/- | T.Shs. 75/- | |
| <u>Material input</u> | | | | |
| Seeds | T.Shs. 95/- | T.Shs. 25/- | - | |
| Fertilizer | - | - | - | |
| Manure | Rare | - | - | |
| Pesticides | - | - | - | |
| Depreciation for implements | - | - | - | |
| Gross margin | T.Shs. 680/- | T.Shs. 110/- | T.Shs. 75/- | T.Shs. 865/- |
| <u>Labor input</u> (in days) | | | | |
| Field preparation (hoeing) | 70 | - | - | |
| Planting | 10 | 2.5 | 10 | |
| Weeding | 60 | - | 10 | |
| Mulching | - | - | - | |
| Fertilizing | - | - | - | |
| Spraying | - | - | - | |
| Pruning | - | - | - | |
| Harvesting | 13 | 8 | 5 | |
| Workdays | 153 | 10.5 | 25 | 188.5 workday |
| Gross margin per workday | T.Shs. 4/44 | T.Shs. 10/48 | T.Shs. 3/- | T.Shs. 4/5 |

In the dry year of 1975, the corn yield fluctuated between 300 and 1,000 kilograms per hectare, concentrating around 750 to 800 kilograms. The wide range was due to the varying intensity with which secondary crops were grown. Corn is normally cultivated in the long rainy season, but no significant difference between the yield for this and the short rainy season could be discerned. The more decisive factor for the size of the yield was the actual quantity of rainfall. Experts estimated the average corn yield in the Usambara mountains to be 750 kilograms per hectare given traditional cultivation, good climatic conditions, and mulching. This yield can be tripled with intensive care, specifically the use of artificial fertilizer. It can be quadrupled by using hybrid seed with the same cultivation techniques plus good irrigation. To achieve this, however, 500 to 700 kilograms of fertilizer per hectare (more than T.Shs. 400/-) are required, which is too expensive for the peasant.

Investments were made mainly for seeds, although some types were occasionally also cultivated by the farmer himself. Artificial fertilizer was not used at all and cow dung only very rarely. The use of pesticides to protect corn was uncommon. In all, things were handled with very little capital input. Labor input, however, was considerable. Land preparation and weeding took a great deal of time, though the secondary products also benefitted.

Gross margins per workday were modest because of the lack of rain. They fluctuated between T.Shs. 1/20 and T.Shs. 8/-. The return on corn, however, was enhanced by the secondary crops. The secondary crops, the most important of which here were beans, needed very little additional expenditure of labor and practically no capital investment, so they increased the gross margin per hectare per workday.

The calculation of the budget for corn is based largely on the appraisal of the crop at market prices. This is unavoidable for the sake of comparability, but it is of only limited value as a criterion of efficiency for a product that is primarily grown for home consumption. Subsistence farming must be understood primarily as the source of the population's food, which, in the opinion of these farmers, is their most important primary need. In good years, however, surpluses may be sold in the market. With regard to corn, subsistence farming is a mode of production which occasionally offers opportunities to make a profit, but which also minimizes market and capital risks because of its minimal monetary input. This does not negate the climatic risks, though. Suggestions for improvement must concentrate primarily on irrigation.

Cassava, bananas. To end this list, cassava and bananas were the most important secondary crops (see table 5.20) and were grown along with coffee in four fields which ranged in size from 1,700 to 3,400 square meters and which were distributed in four of the five subareas. Cassava is very significant as a famine crop, particularly during a drought. Like cassava, bananas are served as alternative food to cornmeal. Neither plant requires much attention, which raises the yield per workday. Both crops are obtained from mother

TABLE 5.20 -

CALCULATION FOR CASSAVA AND BANANAS ON A PEASANT FAMILY FARM

| Field of 1 Hectare | Cassava | Bananas | Total |
|---|---|---|---|
| | Year-round | Long Rainy Season | |
| Yield | 1,500 roots | 200 bunches | |
| Price per unit | T.Shs. 1/- per root | T.Shs. 4/- per bunch | |
| Gross return | T.Shs. 1,500/- | T.Shs. 800/- | |
| Material input | | | |
| Seeds | - | - | |
| Fertilizer | - | - | |
| Manure | - | Some | |
| Pesticides | - | - | |
| Gross margin | T.Shs. 1,500/- | T.Shs. 800/- | T.Shs. 5,250/- (T.Shs. 2,300/-)[a] |
| Labor input (in days) | | | |
| Field preparation (hoeing) | 30 | - | |
| Planting | 30 | - | |
| Weeding | 100 | - | |
| Mulching | - | - | |
| Fertilizing | - | - | |
| Spraying | - | - | |
| Pruning | 3 | 6 | |
| Harvesting | 20 | 15 | |
| Workdays | 183 | 21 | 395 workdays (204 days)[a] |
| Gross margin per workday | T.Shs. 8/20 | T.Shs. 38/10 | T.Shs. 13/29 (T.Shs. 11/28)[a] |

[a] Excluding coffee

plants, with the exception of the banana saplings occasionally bought for T.Shs. -/10 and planted with manure when available. The labor input for cassava far exceeded that of bananas. However, the figure quoted for weeding was somewhat high because weeding benefits the entire field. As with coffee plants, depreciation of the banana plants as perennial crops was omitted.

The gross margins calculated for cassava in relation to the land used, for bananas in relation to the workday, were extremely high. As with corn, however, only surpluses of the risk-minimzing subsistence crops were sold. Their importance lies less in the maximization of profits than in their potential to meet needs, here particularly as a supplementary or additional source of food.

An overview on the various gross returns, some of which still contain many estimates, is given in table 5.21. On this basis the following section combines primary and secondary crops in a mixed farm budget, which is more consistent with the actual situation.

## Farm Budgets

The previous section attempted to produce crop budgets by selecting the most important farm crops paradigmatically. This approach requires monocultural farming or its simulation. Yet this was precisely the method of farming that was neither practiced nor ecologically desirable. Economically, mixed cropping proved at least as profitable if not more so when comparable production techniques were

TABLE 5.21

GROSS MARGINS PER HECTARE PER WORKDAY FOR THE PRIMARY CROPS (IN T.SHS.)

| Crop | Type of Farm | | | |
|---|---|---|---|---|
| | Peasant Farm | | Professionally Operated Farm | |
| | Gross Margin per Hectare | Gross Margin per Workday | Gross Margin per Hectare | Gross Margin per Workday |
| Coffee | 750/- to 3,000/- | 9/- | 6,000/- to 11,000/- | 9/29 |
| Tea | 2,200/- | - | - | - |
| Tomatoes | 2,420/- | 9/49 | 42,655/- | 72/17 |
| Cabbage | 780/- | 3/12 | 2,125/- | 4/49 |
| Corn | 680/- | 4/44 | - | - |
| Cassava | 1,500/- | 8/20 | - | - |
| Bananas | 300/- | 38/10 | - | - |

involved.

In fact, each crop budget must also take secondary crops into consideration. A ceteris paribus clause cannot be used because inputs, particularly of labor, can indeed be assessed for the area of a field, but it is hardly possible to assess them for the individual crop.

Yields from intercropped fields, in turn, can be calculated only by converting them into monetary units. There thus seems to be no meaningful way to break the budgets down according to the individual crops. A detailed reason for this is to be found in the farming method itself. Crops that are complimentary with respect to nutrient requirements and areal expansion--like the combination of bananas, corn, and beans--would take up more space if separated by using a mono-cropping system and would tend to produce smaller yields with commensurately lower returns per unit area.

This shows that the above crop budgets are in fact incomplete farm budgets. The important thing in a farmer's production unit is the *combination* of crops, their potential yield, and their distribution of risks. The analytic but hardly representative calculation based on individual crops, the isolation of which must be simulated, is less important. This leads to the necessity of compiling farm budgets, which are to be used with corresponding risk considerations in formulating production strategies.

The comparison of costs and returns between farms in the following analysis must consider the limiting production factors--land, labor, and capital. The enormous superiority of the professional method of growing cash and export crops is evident in relation to the use of scarce land (table 5.22). Within the peasant farmer's production method, cash crops have certain limited income advantages over subsistence crops. This comparison is meaningless for professionally run farms, where extra food is either bought out of the returns on the cash crops or is produced for personal consumption in a non-professional manner.

The result is less clear in the comparison of returns per workday (see table 5.22). While the index for professionally grown tomatoes is four times that of those that are peasant-grown (the same is true for other vegetables to a far lesser extent), it is lower for coffee. In one case, the immense increase in yield more than paid for the extra expenditure of labor needed, while this was not true in the other case. With subsistence crops like bananas, cassava, yams, or cocoyams, which require little care in the first place, there can be extremely high returns per workday in individual cases even when compared to professional farming methods.

Comparing labor intensity (workdays per hectare per year, in table 5.23) makes this clear. Labor intensity increases across the board in the transition from peasant farming methods to professional cultivation. Since labor, like land, is a scarce input factor for the family farm, the decision about production strategies is influenced not only by the absolute yield or (in monetary units) the return but also by their relation to the production factors of labor, and, as described below, capital.

Capital intensity (the costs of inputs per hectare

TABLE 5.22

RETURNS FOR THE MOST IMPORTANT COMBINATIONS OF CROPS (IN T.SHS.)

| Primary Crops | Peasant Farming Methods | | | Professional Farming Methods | | |
|---|---|---|---|---|---|---|
| | Secondary Crops | Returns per Hectare | Returns per Workday | Secondary Crops | Returns per Hectare | Returns per Workday |
| Tea | None | 2,200/- | - | None | - | - |
| Coffee | Corn, bananas cassava, cocoyams | 1,500/- to 3,800/- | 10/25 | None | 6,000/- to 11,000/- | 9/29 |
| Tomatoes | White cabbage | 3,200/- | 6/34 | White cabbage | 44,780/- | 42/09 |
| Corn | Beans, cocoyams | 865/- | 4/59 | None | - | - |
| Cassava | Bananas | 2,300/- | 11/28 | None | - | - |

TABLE 5.23

LABOR INTENSITY: WORKDAYS PER HECTARE PER YEAR

| Primary Crop | Secondary Crops with Peasant Farming Methods | Peasant Cultivation | Professional Cultivation |
|---|---|---|---|
| Coffee | Corn, bananas, cassva, yams, cocoyams | 125 | 645 |
| Tomatoes | Cabbage | 505 | 1,046 |
| Corn | Beans, cocoyams | 188 | - |
| Cassava | Bananas | 204 | - |

TABLE 5.24

CAPITAL INTENSITY: INPUT COSTS PER HECTARE PER YEAR (IN T.SHS.)

| Primary Crop | Secondary Crops with Peasant Farming Methods | Peasant Cultivation | Professional Cultivation |
|---|---|---|---|
| Coffee | Corn, bananas, cassava, yams, cocoyams | 3/- without secondary crops | 1,419/- without investment for saplings |
| Tomatoes | White cabbage | 220/- | 4,120/- |
| Corn | Beans, cocoyams | 120/- | - |
| Cassava | Bananas | 0/- | - |

per year in table 5.24) take into consideration the running costs, but not the depreciation of investments, which, as with the perennial coffee or banana plants, were made years earlier. This shows that the running investments soar by hundreds of percentage points, even more than labor inputs, with the transition to professional farming methods. Such a leap is possible for family farms only with the support of state aid--in the form of credits or the inclusion in a cooperative system.

It is easy to infer what this means for risk expectation. Where capital investments are involved, the weather-related crop failures not only cost the labor that was lost but also means a cash loss or debt that is huge in comparison with the available income. Debts, in turn, create dependence --and as the need interviews have shown, independence is precisely the motive for choosing the farmer's occupation in the study area.

But even assuming stable weather conditions, the risk inherent in dealing with market prices increases with professional farming methods. Genuine subsistence crops are independent of market prices since they are consumed, not sold. They satisfy the primary need for food by means of direct supply at zero price risk without having to go through exchange procedures or a money economy. If capital input does not exceed zero, as is often the case with the peasant's methods of cultivation, then the risk increases linearly with falling prices when exchange or selling is involved. The greater the proportion of capital in the overall factor input picture, the greater the price elasticity of the gross margins or profits. The price risk or capital risk increases at a disproportionately high rate in relation to falling prices (see table 5.25).

To summarize, the transition from subsistence farming to cash crop farming--and more so from peasant farm methods of production to professional methods--offers the opportunity to increase income, but it also harbors considerable risks such as increased labor input of the extended family, capital risks that are linked to prices, and socioeconomic dependence on the lender when obtaining credit. The opportunities must be weighed against the risks when deciding on a change of production strategy. In the technological domain, it is of utmost importance to seek the methods of farming that reduce the risks in comparison to the opportunities--in other words, those which facilitate an increase in the yield with relatively little additional capital and labor.

## Production Strategies

Another survey obtained additional information about actual production strategies and their motives (see the questionnaire on "Inputs and Outputs," appendix ). Above all, the responses to the questions about the possibilities for enlarging farms or about a fictitious pressure to reduce farm size were intended to provide insights into certain production priorities. This assumes that the marginal considerations (questions 06 to 08) make it possible to discern the maximization of marginal benefit in the decision-making process. The unstructured response items concerning the

TABLE 5.25

RISK EXPECTANCY: PRICE ELASTICITY OF THE GROSS MARGIN (IN PERCENTAGES)

| Primary Crop | Peasant Cultivation | Professional Cultivation |
|---|---|---|
| Coffee | 10.04 | 29.13 |
| Tomatoes | 10.41 | 10.55 |
| White cabbage | 11.54 | 18.35 |
| Corn | 13.97 | - |
| Beans | 12.27 | - |
| Cocoyams | 10.00 | - |
| Cassava | 10.00 | - |
| Bananas | 10.00 | - |

NOTE: The gross margins for the different production methods change by x % if the producer prices change by 10%.

TABLE 5.26

MARGINAL PRODUCTION PRIORITIES (IN PERCENTAGES)

| | Cash Crops | Subsistence Crops | Combination of Both | No Answer | Total |
|---|---|---|---|---|---|
| 1 acre more | 15 | 46 | 39 | 0 | 100 |
| 1 acre less | 31 | 31 | 23 | 15 | 100 |

reasons for the decision made in each case (questions 07 to 09) were intended to reveal the production motives of the farmers so as to verify or falsify corresponding hypotheses. Two final items (questions 10 and 11) focused particularly on the motive of profit maximization. Is the decision based on marginal variations always in favor of the most profitable cash crop? What were the reasons behind a decision that opts for other production strategies? To evaluate the responses, all crops grown must be considered along with the size of the farm, for the hypothesis is plausible that the cultivation of a larger area of land satisfies subsistence needs and thus increases the willingness to accept the risk in the hope of achieving a greater cash income.

If given more land (question 06), 85 percent of those farmers interviewed indicated that they would grow crops exclusively or at least partially for their own personal consumption. Only 15 percent responded that they would grow market crops to the exclusion of all others (table 5.26). The highly profitable crops of coffee and tomatoes were cited here. An analysis of this 15 percent shows that the available land--at 6,000 square meters per family--exceeded the average amount of land that Usambara farmers had. It also shows that an optimal supply was already ensured with less land by growing the most important crops like corn, bananas, cassava, as well as coffee and tomatoes.

The most important reasons given (question 07) for the selection of the additional crops were

- scarcity of food, more food for the family (corn, beans);
- short growing season, low labor input (beans);

- long term cash income for items like clothing (coffee);
- security for the future (cassava);
- reduction of risk in the event that one of the cultivated crops were to fail (simultaneous cultivation of cash and subsistence crops).

The responses to question 08, about production priorities if the farmer were to lose one field, complement the pattern established by these results. One-third fewer farmers were prepared to give up a field devoted to personal consumption needs. Twice as many of those interviewed would rather have produced fewer cash crops and forfeited cash income than threaten their immediate livelihood. Fifteen percent of those interviewed were either not willing or not able to answer because they feared expropriation or were incapable of understanding the abstract sense of the question as a model for discussion.

Additional information on production strategies and their motivations is given in the breakdown of the responses to question 09. Just under one-third of those interviewed opted to give up subsistence crops. However, in no case did this include a staple food. The farmers said they would do without beans, bananas, and sweet potatoes, explaining that these were only secondary crops. The fourth crop that could be forfeited in this context was cassava since its growing period lasts several years and the growing of other crops would ensure a better use of the land that would remain after having given up one field. These responses can be interpreted to mean that short-term supply of food has priority over variety in the diet and long term security.

Cash crops were overwhelmingly "voted down" because the immediate supply of food for the family was considered more important.

Just under one-fourth of those interviewed would prefer giving up a combination of their cash and subsistence crops. The reason given every time was that it was better to give up a little of each crop so as not to change the existing production structure. This means that a combined production strategy was perceived in the cited cases as optimal and should therefore be preserved. These farms had an above-average variety of crops that included not only the typical cash and subsistence crops, but also a variety of secondary produce like tree crops. This leads to the conclusion that the highest satisfaction is achieved with a relative maximum of subsistence along with activities that bring in a cash income. If some things must be given up, however, the above responses indicate that satisfaction of subsistence needs has priority over the satisfaction of needs that can be achieved only by a cash income.

Tomatoes and coffee were realistically cited (question 10) as the most profitable crops by 75 interviewees (58 percent). White cabbage was listed by 25 percent. In answering question 11, two-thirds of those interviewed limited the amount of these profitable crops only because an increase in their cultivation was hampered by scarce factors of production. More than half the responses (54 percent) indicated that farmland was the most important scarce factor. The remaining responses were divided evenly between high risk (including certain doubts that a type of plant will not grow or that a new type has not yet proven itself) and lack of work time and capital. A quarter of the interviewees reported either that they had already optimized their proportion of cash crops or that they were planning to increase it. No one's response to this question indicated willingness to forfeit the increased cultivation of their most profitable crops in favor of growing more food crops. Presumably, the most important type of response explains this: Production factors are necessarily scarce under the unavoidable precondition of having to produce for subsistence needs. The same must also be true of the criteria for optimizing the cultivation of cash crops.

This confirms that an increase in income, particularly the cash portion, is certainly desirable but is in large part not achieved because of external limitations. Assuming that the total amount of available land for farming will not change due to the overpopulation already evident, new farming techniques must be developed that increase labor productivity while capital input rises subproportionately. The risks should be handled by selecting appropriate, ecologically stable varieties of crops.

In summary, the production strategies pursued have integrated an improvement of income into their catalogue of goals but do this under the extremely restrictive conditions imposed by securing the means of subsistence and by balancing risks. Securing the means of subsistence is perceived as the top priority by the majority, which apparently articulates "need"--the needs for food and social security. High risk fluctuations of market prices aside, an increase in income can essentially be achieved only by improving the yield. And proceeding from the limitation of production factors available to the individual, this is itself possible only with the help of more up-to-date, less capital intensive methods of farming.

# CHAPTER VI

# YIELD INCREASE AND ENVIRONMENTAL COMPATIBILITY THROUGH TECHNOLOGICAL PROGRESS

## Perception of Yield Increase

To expand on the assessment made by the observer and to prepare innovative measures by the competent authorities, it seems important to become acquainted with the appraisal of the potential for increasing agricultural production by the farmers involved. This was the intention of an additional survey that submitted for evaluation the catalogue of technical improvements that could be introduced for this purpose (see appendix ). Specifically, the following matters were examined:

- What were the main farming or agricultural problems in the study area and which solutions were or were to be tried (items 02-05)?
- Dovetailing the results on the study of needs was the question: How intensive was the increase in production being pursued and for what reason (item 06)? The question as to how realistic such increases appeared to the interviewees (question 06.2) brought up the assessment of methods.
- The assessment of individual measures should give information about which agricultural techniques the interviewed farmers believed could contribute to an increase in production (item 07). The questions presented in this item

should provide insight into the level of agricultural expertise. How intensive the accompanying information campaign must be when introducing a package of agricultural methods, for example, depends on the responses to such questions. It would focus in detail on the improvement of labor and capital resources and on technological innovations. In the area of technologies, the purpose was to assess measures for mechanization and irrigation as well as the essential aspects of ecologically sound cultivation.

- Finally, with this range of methods in mind, the interviewee was to explain as fully as possible which agricultural techniques he thought were the best and to justify his opinion (items 08-10).

The questionnaire was discussed with thirty-one interviewees in ten groups after being pretested by experts. In the opinion of the experts, the agricultural problems of the investigation area (item 02) were

- drought and sorcery
- infestation with parasites
- soil erosion
- lack of capital
- ineffectual extension service
- inadequate transport.

The experts described the problems very much as the farmers saw them. The frequency of the responses indicated three main problem areas that impede the increase of agriculural production: drought, lack of capital, a decrease in soil fertility, and erosion. Accordingly, the problems were climatic, socio-economic, and, when environmental influences were considered, agro-economic (see table 6.1).

TABLE 6.1

PERCEIVED AGRICULTURAL PROBLEMS
(IN PERCENT OF THOSE INTERVIEWED)

| | |
|---|---|
| Drought | 84 |
| Lack of capital | 74 |
| Loss of soil fertility: erosion | 61 |
| Infestation with parasites | 35 |
| Scarcity of land | 23 |
| Transport | 19 |
| Seeds | 13 |

The question about attempts to implement concrete remedial measures (item 04) revealed a certain helplessness in dealing with the problems. No specific agricultural suggestions were made. The consultations with the rainmaker that undoubtedly occurred in several cases were not mentioned. Only emigration or the search for paid work were reported as solutions in a rather fatalistic way.

In only four of the ten discussion groups was aid from outside sources actually received (item 05), specifically, in three villages cooperating with agricultural extension services. Family or friends were reported as supporters in only one group. Accordingly, four discussion groups could not imagine that outside assistance was possible. The other groups cited the following organizations and measures:

- The government: credit
- LIDEP: transportation
- Incorporation into a cooperative: access to state aid
- Rainmaker: irrigation.

The desire for an increase in production (item 06), which was added as a control question, was shared by all those interviewed. The reason was based as much on food as on income. In all, an increase in production was supposed to provide a better life, or at least keep pace with the increase in living costs. The *possibility* of increasing yield, however, was seen as realistic by only two-thirds of those interviewed. Asked for a reason, the optimists suggested the use of certain production factors or the implementation of other measures such as digging irrigation channels; using manure, fertilizer, pesticides, and better seed; and being more careful in general, such as by regularly pruning the plants. The pessimists said the scarcity of land and the lack of rain and money could not be overcome.

A breakdown of questions into detailed measures (items 07.1-07.5.7) showed that irrigation schemes and all types of fertilizers in addition to chemical pest control were considered to be promising. So was most of the package of methods included in the ecological farming approach. Traditional measures such as crop rotation, temporary fallowing, and mulching were acknowledged. The peasants were equally receptive to the suggestion of planting strips of deep rooting Guatemala grass at regular intervals to prevent erosion. Interestingly enough, some of the interviewees disapproved of intercropping and almost all of them repudiated their practice of tolerating weeds--both measures being traditional. It is still an open question whether this response was intended to prove modernity in order to please the interviewer, whether that person be from the Ministry of Agriculture or from far-off Europe. Every type of mechanization was felt to be just as inadequate since the plots of land to be farmed were too steep and too small. Expending more labor, whether one's own or someone else's, was felt to be more or less futile under the circumstances (see table 6.2).

In a final assessment (items 08-10). the groups discussed the "best" agricultural methods. Articulating an integrated perspective, one group maintained that it was important to care for and work the land well. That included mulching and the process of composting dry material. In addition, they argued that fertilizers and sprays must be used regularly. High yield crop varieties were considered to be essential. Regular applications of fertilizer with equal amounts of manure, and mulching were the most frequent response, followed by the quality of the seeds used, the application of pesticides, and the loosening of the soil. These suggestions were based on personal experience and on the recommendations of the agricultural extension service.

To summarize, the attitude towards the measures stemmed partly from the farmers' personal awareness of problems and partly from the agricultural methods they have practiced. The judgements they made were well-founded, and were consistent with the environmental and socio-economic conditions of the study area. The implementation of measures or programs that deviate from or even contradict the expertise

TABLE 6.2

POSSIBILITIES FOR INCREASING AGRICULTURAL PRODUCTION

| Factor of Production | Possibility for Increase | Prospects for Success[a] |
|---|---|---|
| Labor input | Increase in one's own work | 55 |
| | Use of hired workers | 26 |
| Material input | Improved seed | 90 |
| | Manure | 100 |
| | Fertilizer | 87 |
| | Pesticides | 100 |
| Mechanization | Plow | 26 |
| | Draft oxen | 16 |
| | Tractor | 6 |
| Irrigation | Water bucket | 71 |
| | Irrigation canals | 90 |
| Ecological farming methods | Shade trees | 68 |
| | Horizontal anti-erosion strips of grass | 87 |
| | Mixed cultivation, inter-cropping | 58 |
| | Crop rotation | 100 |
| | Mulching | 100 |
| | Reduced weeding | 10 |
| | Leaving land fallow | 100 |

[a] Percentage of the interviewed peasants who judged the measures likely to succeed.

of the peasants would not only be resisted but would, if past experience is an indication, proceed on inherently false or unrealistic assumptions.

## Approaches to Improving Production

A variety of measures for improving production and for increasing productivity was generally accepted as necessary and need not be discussed further. This would primarily include irrigation and the use of higher-yield, but resistant, crop varieties. Pest control and the use of fertilizers must be improved unless this is already linked to the introduction of ecological measures. To a certain extent, these steps are the basic conditions, the *ceteris paribus* variables.

Moreover, there should be an examination of the few less obvious suggestions that this study indicated might nevertheless be purposeful and worthwhile. Such measures primarily concern crop innovations. They help to enhance the productivity per unit of capital. Still more important are the questions about mechanization since they directly affect the increase of labor productivity. Thirdly, there are at the center of these considerations the methods that increase soil productivity.

### Crop Innovations

Because the population of Lushoto had a rigid scale of preference for its essential subsistence foods and because the cash and export crops of tea, vegetables, and coffee had a secure place,[1] ensuring economic stability in this context must depend only on a few additional crops. From the wide range of Lushoto's less familiar yet more cultivable and worthwhile crops, three have already completed a successful trial period wine, macadamia, and cardamom and are to undergo a cautious economic assessment below.

Wine. Viticulture and winery are not new to the tropics. The population of the Dodoma region in central Tanzania has already had experience with both. The Bihawana mission harvested the first grapes in 1953. Ever since, wine has been produced primarily for personal consumption and for the Catholic Church and its missions in the country. This success encouraged the state support of viticulture in the same region. In 1969 a winery was built in the vicinity of the capital, Dodoma, with Italian aid while the training program was conducted by Yugoslavia. There were two harvests a year. Glucose was added during the process of fermentation in order to produce an alcohol content of 12.5 degrees. After two years in storage, the wine was bottled. Robust red, dry rose, and white dessert wines were produced. The vineyards expanded from 65 hectares in 1969 to 262 hectares in 1976.[2]

Problems stemmed from the natural environmental conditions. Support stakes for the vines had to be transported over 130 kilometers. Because of the dangers posed by the

---

[1]Tea was justifiably in particular demand in official economic planning, and the land area being used for it was being expanded.

[2]W. Kalela, personal communication, November, 1976.

climate and by erosion, an effort must be made to replace the trees after deforestation. Transport problems are incurred by the often irregular harvests in the villages. The limiting production factor, however, was lack of water. Without irrigation, no cultivation at all was possible.

In spite of the undeniable environmental problems and other difficulties associated with viticulture, storage, and marketing, it was evident that cultivation and production of wine was possible and that measures to improve quality were already being implemented. As early as 1976, the quantity (the total yield of two harvests) was 866,707 metric tons of grapes, which represented an average yield of about 5,400 kilograms per hectare for both harvests combined and a total return of just under T.Shs. 2.7 million. The price per kilogram fluctuated betwen T.Shs. 2/50 and 3/50 (Becker and Kiefer, 1977).

A transfer of viticulture to the mountain region of the Usambaras has already been successful.[1] The farming and climatic conditions there are more similar to those in Europe than those in Dodoma. There is only one harvest in February and March, following the hot months. Since the product is of good quality, it appeals to the tastes of the city population and can save foreign currency by substituting for imports.

---

[1]Brother Mathias, Sakarani Mission, personal communication, October, 1976.

Macadamia. The madamia nut comes from Australia (where it is also called the Queensland Nut). It is spherical with a diameter of about one centimeter, and has an extremely hard shell. The white kernel tastes similar to the hazel nut. When roasted and salted, the macadamia is considered to be the most expensive nut in the world. After sixteen years one tree up to twenty meters tall and thirteen meters in diameter produces ninety kilograms annually, and 135 kilograms after thirty-two years. The oil content of the kernel ranges 67 percent to 72 percent (Grimwood, 1971; Acland, 1973).

The macadamia adapts well to other tropical and subtropical regions. In the Kenyian highlands it has become a diversification crop and shade tree in coffee fields (Glaeser, 1975 p. 80). Its cultivation was planned for Tanzania's southern mountain regions like Mbeya or Iringa. A few trees had already been planted, with good results, in the Usambara mountains.[1]

The tree lends itself, for example, to planting in the school gardens, for the nut's high protein content improves the children's diets. Corn can be intercropped with the young trees. Small-scale industrial processing is also conceivable, which could create additional rural jobs. The nut's export potential should not be overlooked, either, provided that a certain domestic demand offsets the risk of overgrowing and processing the product.

---

[1] Brother Mathias, Sakarani Mission, personal communication, October, 1976.

Cardamom. Experimental growing of cardamom and other aromatic plants was successfully carried out by the German experimental station in Amani in the eastern Usambaras as early as the turn of the century. Since then, cardamom has developed into an important export product of the Tanga region, particularly for the peasant.

Cardamom grows best in the highlands, especially in former forest areas. In addition to the quality of care, which includes intensive field preparation, mulching, weeding, drying and storing are also decisive in determining the level of returns. In 1975, one kilogram of cardamom fetched T.Shs. 12/-, and T.Shs. 28/- when sun-dried. The yields in full maturity come to 125 kilograms per hectare, the gross profits to T.Shs. 380/- per hectare or T.Shs. 23/- per workday. With improved care, drying, and storing, T.Shs. 700/- per hectare are possible, though only at gross profits of T.Shs. 17/- per workday (Tanga Plan, vol. 2, pp. 85-86 and vol. 3, pp. 270-271).

For farmers of the western Usambaras whose subsistence level has already been met, cardamom is thus to be recommended as an additional export product to bolster income. The land area used for this should be expanded cautiously, however, since this crop's uncertain situation on the world markets requires a willingness and an ability to assume great risk. Larger growing areas can also upset the ecological balance. Soil erosion and water problems are often symptoms of excessive deforestation.

## Mechanization

Ecological farming procedures that are appropriate for tropical conditions are one possibility for increasing agricultural production without eventually destroying its ecological foundation. This does not preclude the possibility that more efficient use of human labor might also increase production. These considerations bring up several questions. Which forms of mechanization exist? Which forms can be introduced spontaneously? Which of these forms are consistent with the region's established production patterns and do not threaten the long-term preservation of soil fertility? To answer these questions, it may be expedient to offer a brief overview of the tools and implements in Tanzania itself, as they must be readily available, relatively inexpensive, and must not cost too much money--especially foreign currency. For the present purposes of improving agricultural output, it seems appropriate to categorize the country's existing forms of mechanization according to their function in farming.[1]

- Preparing the land
- Working the field
- Weeding
- Harvesting
- Processing the harvest
- Transport

---

[1] The author draws here on illustrative material obtained during visits to artisans' workshops run by the Tanzania Agricultural Machinery Testing Unit (TAMTU) in Arusha and LIDEP in Soni. Supplementary information is found in the

Preparing the Land. The hand hoe (*jembe*) and the machete-like *panga* are still the only instruments used on peasant farms. One could improve both by adapting them to the type of work or by using animal power. The first suggestion could involve the use of more durable material. The blade of the hoe, for instance, could be fashioned from more highly tempered steel. Constructive improvements would have to consider various handle lengths, blade widths, grips, and angles of incidence. Single or multiple-pronged plows that can be pulled by oxen have already been developed by TAMTU. The three-pronged plow cost T.Shs. 915/- in 1974. An animal-drawn disc harrow, also developed by TAMTU, costs T.Shs. 980/-. It is doubtful, however, whether teams of oxen can be used effectively in the fields situated on the steeper slopes.

Working the Field. Traditionally, the fields are worked in a very labor-intensive way. A hole is dug, and the seed is placed into it by hand. Hand and animal-drawn implements are widely unknown, although they would save both time and seed. Favorable sowing schedules could be observed more regularly, too.

The same argument against the general use of ox-drawn implements can be used against TAMTU's hand planter developed for the three-pronged plow for T.Shs. 200/- (1974). Hand implements like the one developed by LIDEP and sold for T.Shs. 45/- in 1976 are certainly easier to use here. The tip of this tool is pushed into the ground whereupon a

---

reports by Macpherson (1975), pp. 116-185 and Tschiersch (1975), pp. 94-125.

mechanism releases the seed, which then falls into the soil. This eliminates tedious bending. Labor efficiency soars to between one-half and one hectare of land completed per day. Similar implements, some of which are even simpler (without spring mechanisms), were developed both by TAMTU's Department for Village Technology and the Department of Rural Engineering at the University of Dar es Salaam for sowing corn and beans. The introduction of such tools is worthwhile when there is a lack of labor or when seed prices are high. There are no topographical difficulties. An additional advantage is that subsequent weeding can be done much more quickly if the seeds are carefully sown in rows, which can be assured by using a guide string.

Weeding. As with preparing the land, one-hand hoes are used for weeding. This chore takes more time than any other. TAMTU's price for a cultivator (a weed plow whose width is adjustable) was T.Shs. 280/- in 1974. This implement can be pulled by oxen or donkeys. Because the LIDEP's donkey program in the Usambaras was not successful, however, the use of this tool can not be realistically considered.

Harvesting. Nonmotorized implements for mechanizing harvest work are not available. But the labor is not a great limiting factor in this functional domain, anyway, because harvesting can be distributed over a long period of time unless the next sowing deadline is pressing.

Processing the Harvest. The shelling and milling of corn is the most important activity in processing the harvest. Pounding with mortar and pestle is still seen in the villages. The advantage--the preservation of vitamins and other nutrients--is kept with simple hand milling, though. A hand-powered maize sheller suitable for household use was sold by TAMTU and LIDEP for T.Shs. 80/-. There were numerous motor-driven corn mills run as small businesses or as a communal village activity.

Transport. Traditionally, most travel like the necessary trip to the various markets is done by foot with the goods carried on the head. The widely advertised ox-drawn or double mule-drawn wagon is certainly less suited to the mountains. Experience on Mount Kilimanjaro has shown, however, that the donkey can be used as a beast of burden. Wheel barrows and handcarts have also been utilized and have established themselves in certain areas.

Evaluation. There have been promising beginnings for efficient, appropriate techniques in the agricultural sector, but in general there is far too little technical, financial, and training support. There are fairly few implements that appreciably increase the productivity of rural labor without involving excessive capital expenditure. There is a technological need that should be satisfied through innovation within the domestic workshops that already exist. The use of appropriate mechanization in agriculture could complement the ecological improvements and make them more effective.

## Ecological Techniques

Ecological techniques are central to the ideas and suggestions for increased production combined with resource conservation and environmental protection. They involve a synthesis of traditional African agriculture and European ecological methods (Egger and Glaeser, 1975) that have been modified for tropical conditions (Egger and Mayer, 1976). This section presents the initial results of introducing such a pattern of methods in the Shashui study area during 1975 and 1976.

Experimental Fields. Fields in the village of Msikitini, located near the market town of Soni, were chosen for experimentation which was planned according to the principles of ecological farming. This was because Msikitini was readily accessible with the local road network and, being surrounded by hills, because it comprised a relatively closed ecosystem. The experiment was carried out with the cooperation of the native farmers. First, strips of Guatemala grass were planted in contour rows to control erosion. As a further measure, bananas, cassava, corn, and beans were intercropped in contour rows to facilitate the work, to make for more even irrigation, and, finally, to ensure against erosion. Guatemala grass is suitable for cattle fodder and so provides the basis for a closed system of mixed agriculture. To support all these measures, a small tree and seed nursery was established on the floor of the valley. Only materials accessible to the native peasants--like the support stakes and shady banana leaves--were used. More Guatemala grass was to be grown along with tree crops such as avocados.

The three pilot fields and the seed nursery described below--all developed under the supervision of biologists (Egger and Mayer, 1976)--will first be appraised on the basis of the inputs used and the costs incurred. This will be followed by deriving the production function of hypothetical plots one hectare in size. The result can, especially after projection to the entire study area of Shashui, provide clues about the initial expenses that could be expected if the ecological innovations were to be translated into practice with the help of experts. Such clues are valuable for the local authorities ultimately responsible but give no indication per se of the cost-benefit relation on the peasant farm. A quantitative assessment like that is to be provided at the end of this chapter and coupled with a critical analysis of the pattern of ecological methods. Ideally, a reliable series of experiments should continue over several years, but, failing that, it should at least include a large number of farms. This was not possible given the short duration of the observation period. The appraisal presented in table 6.3 is thus to be checked through further agricultural experiments.

To begin the above described appraisal, table 6.3 presents the total inputs and costs incurred by the development of all three experimental fields and the seed nursery in Msikitini.

TABLE 6.3

ITEMIZATION FOR MATERIAL, TRANSPORT, AND LABOR, 1976 (IN T.SHS.)

| Inputs | Itemization | Costs | |
|---|---|---|---|
| Farmers' supply shop | Fertilizer TSP per kg | 3/40 | |
| | Fertilizer NKP per kg | 3/40 | |
| | Lime per 50 kg | 15/- | |
| | Cement per 50 kg | 20/20 | |
| | Seeds for hybrid corn per kg | 50/50 | |
| | Seeds for beans per kg | 8/- | |
| LIDEP prices for vehicles and labor | Dumptruck (5-ton) per km | 5/- | |
| | Truck (3-ton and 7-ton) per km | 5/- | |
| | 5 tons loaded (unit price) | 10/- | |
| | 7 tons of bricks loaded (-/01 per brick) | 30/- | |
| | | month | day |
| Wages | Casual labor | 250/- | 10/45 |
| | Minimum wage for skilled worker (e.g., driver) | 313/- | 12/50 |
| | Secondary school boy | 410/- | 16/- |
| Guatemala grass | Price for 1 sprout | - | |
| | Transport: 200 sprouts in landrover from Maweni to Shashui (approximately 11 km at 2/30 per km) | 25/- | |
| | Planting: 30 sprouts per workday by a secondary school boy | 16/- | |
| Bananas | Price for 1 tree (-/40 in Mamba; 1/- in Msikitini) | -/40 | |
| | Transport: 100 trees in landrover from Maweni to Shashui (cost based on experience) | 45/- | |
| | Planting: 10-12 banana trees carried from the street to the field; holes dug; composted animal dung fetched; watering; setting tree into hole (1 workday) | 16/- | |

| | | |
|---|---|---|
| Cassava | Price for 1 stalk = 4 cuttings | -/10 |
| | Transport: 500 stalks in landrover from Muheza to Shashui (2 X 120 km at 2/30 per km) | 552/- |
| | Planting: 500 cuttings = 1 workday | 16/- |
| Grevillea, Albizia, Avocado | Price for one sapling | 1/- |
| | Transport: 200 saplings in landrover from Lushoto to Shashui (2 X 15 km at 2/30 per km) | 69/- |
| | Planting: 10 saplings = 1 workday | 16/- |
| Lime from Tanga | Price per ton | 250/- |
| | Transport: 3-ton truck of Maweni limestone near Tanga to Shashui (2 X 150 km at 5/- per km) | 1,500/- |
| | Loading and unloading (each 10/-) | 20/- |
| | With the byproduct of lime powder, the price can drop to 0. The same is true for transport costs if the return trip of the vegetable trucks is calculated as opportunity costs, though this was unlikely under the customary LIDEP accounting system. | |
| Lime from Dar es Salaam | Transport: 5-ton dumptruck or 3-ton or 7-ton truck (2 X 375 km at 5/- per km) | 3,750/- |
| | If the return trip of the LIDEP vegetable trucks is accounted for as opportunity costs, the price may be lower. | |
| Lava from Moshi | Transport: 2 X 220 km at 5/- per km | 2,200/- |
| Phosphate from Lake Manyara | Transport: 2 X 430 km at 5/- per km | 4,300/- |
| Phosphate from Tanga | Transport: 2 X 160 km at 5/- per km | 1,600/- |
| | Only the costs for the 3, 5, or 7-ton vehicles are listed here since they presumably represent the lion's share of the total costs. Unlike possible fertilizers from distant places listed above, using the bark of wattle trees from Lushoto would minimize transport costs, though it is probably still unclear if the bark would induce microbic processes in the first place. | |

In field 1, located on a steep slope without irrigation, the measures presented in table 6.4 were taken. The soil was first loosened, which the owner (the TANU chairman) probably had done by paid workers. Fortification rows were then planted under the supervision of experts

TABLE 6.4

CALCULATION OF COSTS FOR
EXPERIMENTAL FIELD 1 IN MSIKITINI

| Inputs[a] (field size: approx. 30 m X 60 m = 1,800 $m^2$) | Items and Amounts[b] (in T.Shs.) | Costs (in T.Shs.) |
|---|---|---|
| Materials . . . | 180 Guatemala grass sprouts (no charge) | - |
| | 20 bananas at -/40 each | 8/- |
| | 30 pigeon peas (gathered) | - |
| | grevillea at 1/- each | - |
| Subtotal . . | | 8/- |
| Labor . . . . . | 180 Guatemala grass sprouts planted (6 workdays at 16/- each) | 96/- |
| | 20 banana trees planted (4 workdays at 16/- each) | 64/- |
| | 30 pigeon peas (subsistence level) | - |
| | grevillea | - |
| | trimming the Guatemala grass (1/2 hour) | - |
| Subtotal . . | | 160/- |
| Transport . . . | 20 banana trees from Mamba with land-rover (100 at 45/-) | 10/- |
| | 180 Guatemala grass sprouts from Maweni (approx. 10 km; 200 at 25/-) | 22/50 |
| | grevillea from Lushoto | - |
| Subtotal . . | | 32/50 |
| Total Costs of planting field 1[a] | | 200/50 |

[a] Without using European experts.

[b] Six fortification rows of Guatemala grass, pigeon grass, bananas, grevillea at intervals of approximately 10 m.

Field 2, located above Msikitini, on a steep slope without irrigation, was worked by experts. The nonresident owner showed little interest. The measures taken in this field are presented in table 6.5.

TABLE 6.5

CALCULATION OF COSTS FOR
EXPERIMENTAL FIELD 2 IN MSIKITINI

| Inputs[a] (field size: approx. 20 m X 60 m = 600 $m^2$) | Items and Amounts[b] (in T.Shs.) | Costs (in T.Shs.) |
|---|---|---|
| Materials . . . . | 100 Guatemala sprouts (no charge) | - |
| | 11 banana trees at -/40 each | 4/50 |
| | dung (subsistence level) | - |
| Subtotal . . . | | 4/50 |
| Labor . . . . . . | 11 banana trees carried from the street to the field; holes for plants dug; composted manure fetched from Ntuii (only a few hundred meters away); watering and planting comes to 2 workers for one day (2 workers at 16/- each) | 32/- |
| | 100 Guatemala sprouts planted (3 workdays at 16/- each) | 48/- |
| | pigeon peas planted | - |
| | Guatemala grass trimmed | - |
| Subtotal . . . | | 80/- |
| Transport . . . . | 11 banana trees from Mamba (100 at 45/-) | 5/- |
| | 100 Guatemala sprouts from Maweni (200 at 25/-) | 12/50 |
| Subtotal . . . | | 17/50 |
| Total costs of planting field 2[a] | | 102/- |

[a] Without using European experts.

[b] Four fortification rows of Guatemala grass and bananas at intervals of approximately 8 m.

Field 3, located below the street in the direction of Maweni, was already planted with trees and bananas. The owner showed interest in the measures taken, which are presented in table 6.6.

TABLE 6.6

CALCULATION OF COSTS FOR
EXPERIMENTAL FIELD 3 IN MSIKITINI

| Inputs[a] (field size: approx. 18 m X 30 m = 540 $m^2$) | Items and Amounts[b] (in T.Shs.) | Costs (in T. Shs.) |
|---|---|---|
| Materials . . . | 90 Guatemala grass sprouts (no charge) | - |
| | 25 banana trees at -/40 | 10/- |
| | 30 pigeon peas (subsistence level) | - |
| | 500 cassava cuttings (= 125 stalks at -/10 each) | 12/50 |
| | Seed: sweet potatoes (sprouts, subsistence level; | - |
| | corn (5/50 per kg) 2 handfulls (approx. 250 g); | 1/50 |
| | beans (8/- per kg) 2 handfulls (approx. 250 g); | 2/- |
| Subtotal . . | | 26/- |
| Labor . . . . . | 4 rows of Guatemala sprouts planted (3 workdays at 16/-) | 48/- |
| | 25 banana plants (5 workdays at 16/-) | 80/- |
| | 30 pigeon peas (subistence level) | - |
| | potatoes, corn, beans sown | - |
| | 500 cassava cuttings planted and pruned (unskilled labor, 1 workday) | 10/- |
| Subtotal . . | | 138/- |
| Transport . . . | 90 Guatemala sprouts from Maweni (200 at 25/-) | 12/- |
| | 25 banana plants from Mamba (100 at 45/-) | 11/- |
| | 125 cassava cuttings from Muweza (500 at 552/-) | 138/- |
| Subtotal . . | | 161/- |
| Total costs of planting field 3[a] | | 325/- |

[a] Without using European experts.

[b] Fortification rows at intervals of about 7.5 m using Guatemala grass, pigeon peas, bananas, and cassava at intervals of 1 X 1 m.

On the basis of these calculations for each of the three experimental fields in Msikitini, the production function of a plot one acre or one hectare large can be inferred (see table 6.7).

TABLE 6.7

INPUT STRUCTURE AND COSTS FOR AN IDEAL MODEL FIELD OF 1 ACRE/1 HECTARE

| Inputs | Itemization | Costs (in T.Shs.) |
|---|---|---|
| Guatemala grass | 600 sprouts (no charge) | - |
| | Transport | 80/- |
| | Planting (20 workdays at 16/-) | 320/- |
| | | 400/- |
| Bananas | 80 trees at -/40 each | 32/- |
| | Transport | 40/- |
| | Planting (15 workdays at 16/-) | 240/- |
| | | 312/- |
| Cassava | 4,000 cuttings = 1,000 stalks at -/10 each | 100/- |
| | Transport | 900/- |
| | Planting (8 workdays at 16/-) | 128/- |
| | | 1,128/- |
| Grevillea, Albizia, Avocado | 25 saplings at 1/- each | 25/- |
| | Transport | 9/- |
| | Planting (2.5 workdays at 16/-) | 40/- |
| | | 74/- |
| Costs for 1 acre model field | | 1,914/- |
| Costs for 1 hectare model field[a] | | 4,785/- |

NOTE: The calculations pertain to an easily accessible location in Shashui.

[a] 1 acre = approximately 4,000 $m^2$.

Cost Projection for the Measures. In principle it is possible to extrapolate the costs for each village or for the Shashui study area as a whole on the basis of the size given for the plots of cultivated land. The total costs of the required measures can thus be calculated

The entire area of cultivated land was about 500 hectares. This estimate is arrived at realistically by halving the amount of all cultivated fields reported by the contact persons (chapter III). On this basis and in keeping with the type of ecologically sound measures implemented and subsequently evaluated in economic terms (which included provision and planting of Guatemala grass, bananas, cassava, and various trees), a budget of T.Shs. 2.4 million would be required (DM 800,000 at the exchange rate of 1976).

Such a sum may seem utopian considering that entire mountain regions should profit from these types of measures. Even so, it is certainly possible to reduce costs if specific matters are taken into consideration. All pilot projects, for example, include costs of initiating innovations, expenses which can later be eliminated or reduced. Transport of plants can be reduced to one quarter of the costs reported here by using trucks instead of landrovers. Above all, domestic administrative measures should be planned that would appeal to the spirit of self-help by making the materials available only to those farmers or communes that undertook to do the planting themselves. This could be carried out under the supervision of the agricultural advisors, who are responsible for the region anyway. The computation would then be:

| | T.Shs. |
|---|---|
| Plants/acres (as before) | 157/- |
| Transport/acre (25% of landrover costs) | 257/- |
| Labor per acre (eliminated) | - |
| Total per acre | 414/- |
| Total per hectare | 1,035/- |

This reduces the required external financing to T.Shs. 414/- per acre or T.Shs. 1,035 per hectare, which, with 500 hectares of arable land, would require somewhat more than half a million shillings (T.Shs. 517,000).

Even these costs would be valid only for a crash program. Naturally, it would make more sense to promote regional independence along with the measures for cultivation by encouraging the expansion of each region's stock of seed. This should be done on a cooperative basis since few peasants have the land necessary for it. How much is saved depends on the size and number of such seed nurseries. The first attempt to increase the seed stock in Shashui is described above. Economic appraisal is contained in tables 6.8 and 6.9.

<u>School Program</u>. The ecological measures, erosion control, and production improvements described above were disseminated first by the primary schools in the immediate area. The school teachers and the Ministry of Education (*Elimu*) in Dar es Salaam showed interest because an ecological school garden program complies with the Ministry's call for a certain degree of autonomy in the schools. Within the framework of

TABLE 6.8

COSTS OF SEED NURSERY NEAR FIELD 1

| Inputs and Itemization (Field size: 18 m X 1 m = 18 $m^2$) | Costs (in T Shs.) |
|---|---|
| Transport of forest soil with LIDEP dumptruck (20 km at 5/- per km) | 100/- |
| Loading | 10/- |
| 2 x 5 hours labor in the forest at 10/- per hr | 100/- |
| Establishment of the nursery and planting, setting stakes (3 x 4 days = 12 workdays at 16/- per day) | 192/- |
| Banana leaves fetched from Mamba and roof covered (2 x 3 days = 6 workdays at 16/- per day) | 96/- |
| Guatemala grass (54 cuttings at 0/-) | - |
| Avocados and other seeds | - |
| Transport of Guatemala grass (54 sprouts) from Maweni to Shashui | 6/50 |
| Cost of the grounds . . . . . . . . . . . . . | 504/50 |

TABLE 6.9

INPUT STRUCTURE AND COSTS FOR AN IDEAL MODEL GUATEMALA GRASS SEED NURSERY OF 1 ACRE/1 HECTARE

| Input | Itemization | Costs (in T.Shs.) |
|---|---|---|
| Material | 6,000 sprouts per acre | - |
| | (15,000 sprouts per hectare) | |
| Labor | 6,000 sprouts cut and carried to the street (2 workers X 1 workday for 500 sprouts) | |
| | 24 workdays at 10/- per day | 240/- |
| Transport | 6,000 sprouts in landrover from Lukozi to Soni for Shashui (2 X 40 km at 2/30 per km; 500 sprouts per trip): | |
| | 184/- X 12 | 2,208/- |
| Labor | 6,000 sprouts: soil turned up; hole dug; sprouts set in holes; soil packed around roots (2 workers X 1 workday for 500 sprouts) | |
| | 24 workdays at 10/- per day | 240/- |
| | Keeping field clear, weeding | - |

Costs for establishing a model field of Guatemala grass:

1 acre . . . . . . . . . . . . . . . . . 2,688/-

1 hectare . . . . . . . . . . . . . . . 6,641/-

agricultural instruction, pupils and teachers should be in a position to produce the food to meet their needs and possible even to achieve a surplus to finance the teaching materials.

The five primary schools participating in the ecology program in 1976 were located in Soni, Kisiwani, Baghai, Shashui, and Bumbuli. The school garden program embraced the steps outlined below (Kimela, 1976):

- Planting tree crops and banana saplings relatively far apart
- Intercropping with various subsistence crops like corn, beans, and high-return vegetables like potatoes and tomatoes
- Mulching
- Spreading composted manure
- Terracing

Because all five school were located on slopes, each began by digging and building up the one meter wide terraces across the incline. The importance this had for irrigation and erosion control was also stressed in the instructional program. The run-off rain water was channeled from the end of one terrace into the next just as in a canal system. The effectiveness of erosion control can be enhanced by planting deep-rooting Guatemala grass at the edges and on the slopes. The program also included the use of the grass for mulching material and cattle fodder. The cultivation program was finally initiated on this basis and run on a plot of land one-half to four hectares in size. Corn, beans, white cabbage, tomatoes, sweet potatoes, and onions were planted as needed. The limited use of pesticides was taught especially in connection with vegetables.

The example of the Shashui primary school is presented below and evaluated on a quantitative basis as far as the period covered by the investigation allows. The village made available about four hectares of land which did include a fertile valley floor but mostly land of little value on the slopes. The incline was terraced in the manner described above. Corn was planted at the edges and surrounded by rings of mulch in which water collected. Tomatoes and carrots were the main vegetables. Table 6.10 presents the farm budget for one year as given by the teacher responsible.

TABLE 6.10

ESTIMATED FARM BUDGET FOR THE PRIMARY SCHOOL IN SHASHUI (IN T.SHS.)

| | Corn | Vegetables |
|---|---|---|
| Gross returns | 900/- | 1,000/- |
| Seeds | 83/- | 660/- |
| Fertilizer | 140/- | |
| Gross margin | 1,017/- | |

The manure provided by the villagers is not included, nor is the expenditure of labor even roughly calculable in this rudimentary computation. What can be said is that the the work needed for terracing cannot be performed by a peasant family, but it could be accomplished in cooperatives or as a communal activity. Above all, it was important for the people to perceive not only that the ecological program can be carried out, but that it is worth emulating, too. Farmers in Mazukizi, Magila, and Msikitini who were following the curriculum of their children were already applying the principles of the school program to their own fields during the observation phase.

## Evaluation of the Measures for Improving Production and Increasing Yields

The permanent improvement in the peasant's income is the goal, with income being earned in the form of cash and/or means of subsistence. This goal requires the achievement of a consistently high yield over a prolonged period of time. This implies that the fertility of the soil should not be threatened by overcropping and that measures to protect and preserve the environment should be implemented. By so doing, the criterion of harmonizing economy and ecology can be met. A difference between economy and ecology is that the former is short term, the latter long term in perspective. Within the lo term scope promoted by this study, the most important role is played by microeconomic factors, specifically those having to do with production theory. The optimal combination of the production factors to be introduced must be found. In order to minimize costs, identified constraints in the factors of production must be relieved.

Since land is clearly scarce and will remain so because of the high rate of population growth, the productivity of the land must be increased. In a region that is overpopulated and that has few job opportunities, the solution would seem to lie in using more labor, either by extending work time or by recruiting additional workers. But since it has been shown that labor cannot be increased ad infinitum, the challenge is to increase labor productivity along with land productivity. Both are achieved traditionally through hiking capital input in the form of mechanization. But this is precisely where there is a particularly severe constraint among peasants in developing regions. The dilemma unfortunately so typical of underdevelopment emerges: almost all factors are scarce, but the means to eliminate this scarcity do not exist.

That is why technological measures to increase yield and improve production are being sought, measures that will increase soil and labor productivity with the lowest possible increase in capital expenditure. In addition to the various crop innovations that increase cash income, the main ways to achieve these goals are the introduction of limited mechanization appropriate to the ecological and socioeconomic conditions of the respective area and use of a variety of ecological techniques. The combination of these two approaches creates a different kind of production pattern whose main characteristic is that it uses nature and physical environment as production factors and uses ecology as an economic and technical means of production--in both cases more intensely than in traditional farming and chemotechnical agriculture.

Because the experience with ecological measures has thus far been brief, there can and should be only a tentative economic evaluation. Still, erosion control by planting Guatemala grass is a longer term measure whose agricultural advantages will already be observable in the medium term when it prevents the top soil from being eroded and thus preserves soil fertility. Furthermore, growing Guatemala grass provides an additional reservoir of cattle fodder that can be used to full advantage with stall feeding. The proportion of grazing land that is saved in this way would more than compensate for the amount of land lost in the erosion control strips.

Planting crops like bananas or corn in contour rows does increase labor because planting the cross rows demands more work than simply digging the holes for the plants. On the other hand, planting the crops in rows has various ecological and economic advantages such as

- greater density of plants and thereby a greater yield;
- internal rotation (fallow rows) without losing an entire field (minimization of risk);
- controlled weeding and mulching, which makes some of the hoeing unnecessary and therefore saves labor;
- easier cultivation, possibly supported by small-scale mechanized farming methods.

*Intercropping* has successfully been applied in traditional farming in that crops with complementary nutrient requirements are planted together in the same field. A quantitative comparison of yield with monocultures is difficult, but certainly the inputs and costs of fertilizers and pesticides decrease in such an ecologically sound system.

In addition to an improvement in yield, in cash return, and long-term preservation of soil fertility, the optimal combination of production factors is also determined by the chances a crop has of making a profit and the willingness of the farmer to assume risk. The crops grown primarily for home consumption (subsistence farming) tend to reduce farming risk and, with it, of course, the profit opportunities. Vice versa, profit opportunities increase with the cultivation of cash crops, especially those bound for export. At the same time, however, there is a rise not only in risks due to the fluctuation of prices on the world market but usually also to the capital risk linked to the greater volume of inputs.

The decision that regulates the proportion of subsistence production to market production is made according to the needs of the peasant family. Generally, subsistence farming has the highest priority, but the willingness to accept risk increases as personal needs are met. This cannot be automatically assumed for the study area, however.

# Part C

# Conclusions for an agricultural development strategy

# CHAPTER VII

## TRANSFERABILITY OF RESULTS

The relevance of the statements in the preceding chapters depends on the degree to which the results obtained from the specific case example are generalizable and transferable to other areas and regions. The criterion for this is whether the respective socioeconomic and ecological circumstances are comparable.

The ecological problems in the study area are erosion on the slopes, overgrazing, and the scarcity of water linked to deforestation. Socially and economically speaking, the population lives primarily on a subsistence basis. Their problems include scarcity of land, supply difficulties, and lack of money. Education and agricultural expertise are usually of the traditional type or based on experience. The need structure is very strongly oriented to the satisfaction of primary physiological needs.

These critera are also valid for other tropical highlands in East Africa, which are ecologically comparable. In spite of a few cultural differences in the patterns of settlement, for example, all these areas suffer from similar social and economic problems that stem wholly or in part from the high population density. First among these are the problems of erosion on the slopes, deforestation, and lack of ground water, all of which have to do with both ecological and economic factors.

Examples of such highland locations are Mount Kilimanjaro in northern Tanzania, the Kandara region in Kenya near Nairobi, and the mountainous countries of Rwanda and Burundi (Glaeser, 1977a). Rwanda and Burundi are the most threatened countries since they have practically no unused areas. Both countries are among the least developed in the world, are almost purely agricultural states, and have the highest population density in Africa.

The annual population growth rate is 3 percent, which means that the population will double within twenty years. Emigrees are not welcome in the neighboring countries of Uganda, Zaire, and Tanzania because those countries have their own problems. Thus, more and more steep slopes must be cultivated, which is steadily compounding the dangers of erosion caused by rain. The quality of the soil is decreasing unchecked with the zone of marginal yield approaching inexorably. It is no longer possible to maintain food supplies. A single erratic rainy season will lead to famine in the population. Most of the originally abundant forest areas have been cleared, destabilizing the climate even more. Reforestation is possible only to a limited degree because of the population pressure and the need for food plants. As a status symbol, livestock are kept in large numbers, which has resulted in overgrazing and limited milk production.

The cost of "Green Revolution" technology excludes it as a possible solution in this case. Chemical fertilizers and pesticides have become so expensive since 1974 that the individual peasant can no longer afford them. Alternative solutions may be sought in improving traditional agriculture by using ecological methods. Transverse planting on the slopes, mulching, covering the soil, mechanical weeding, strict rotation of crops to protect against pests, and improvement of the soil by intercropping nitrogen-producing legumes like beans have been effective. The most valuable experiences in this regard were gained by Compagnon Batisseur, the Belgian agricultural project begun in 1969, and the equally old German project in Nyabisindu, originally begun for dung production. A similar domestic project was administered by the Anglican Church and financed by World Neighbors.

Rwanda's state planners have pursued approaches to introduce ecological farming methods into general strategies of village development--as shown by the 1977-1981 five-year plan and by talks within the Ministry for Social Affairs and the Rwandan Development Bank. These observations are intended only as brief examples of the transferability to other areas of approaches and results of the work on the Tanzanian case study and will not be pursued further. To conclude, there is the question of the consequences such an approach entails within the framework of the theory and practice of developmental policy.

# CHAPTER VIII

## THEORETICAL IMPLICATONS

Which theoretical conclusions can be drawn from this study? What is their significance for needs orientation, self-reliance, and environmental compatibility--the three elements of ecodevelopment that have been the guiding principles for the empirical execution of this investigation?

### Needs Orientation

The most important result of the needs study is that there was no differentiation between alienated labor and consumption-oriented leisure and that importance was consciously attached to the fact that the three basic physiological needs required the satisfiers of food, clothing, and shelter. These results were arrived at under certain theoretical and technical delimitations.

The needs concept that originated in the theories of Maslow proved to be of little use in conducting the study. This was due in part to the limited range of the concept's theoretical validity, and in part to the lack of a way to operationalize it. In particular, it was found that not only the needs themselves but also their conceptualization are historically and socially conditioned, that the postulated hierarchy in which one need leads to the next highest is very dubious. This ranking may make sense as a *distinctio rationis* just as theoretical hierarchies of goods like Aristotle's can begin to shed light on the problem of needs, but it is virtually useless as a practical research concept.

Rather, it seems more realistic to hypothesize that there is a twin continuum of physical and nonphysical needs whose priorities can alternate according to historical, socio-economic, and geographical influences.

Some of the theoretical inadequacy can be offset by the spontaneity of structured group discussion and through cross-checking needs information with objective data on material requirements. The following implications for future studies of this kind emerge from this appraisal.

- There is no theory of needs for developing countries, but it would be desirable to have one. Empirical studies thus face a dual problem. They have no theory from which to proceed, yet are supposed to contribute to one that is being developed.
- Group discussions in this undefined field are more important than interviews with individuals. Furthermore, unstructured discussions on the topic must be conducted in order to expose any prejudices contained in the methodology or subject matter.
- The information gathered through surveys, dialogues, and discussions should be cross-checked with objective data and supplemented with observations by participants as far as time allows.
- Finally, future investigations that already have a certain theoretical foundation should, of course, satisfy the

methodological standards of empirical sociological research.

## Self-Reliance through Production on the Peasant Farm

Self-reliance--drawing on the human and natural resources indiginous to the country or region--was isolated as the second basic element of ecodevelopment. The realization or nonrealization of this element in non- or underindustrialized areas is decided almost exclusively in the agricultural sector, that is, on the peasant farm (if the agricultural production is disaggregated). For it is the peasant farm that must produce most of the food supply. The main individual production strategies in the study area seek first and foremost to secure subsistence and to balance risks. In macroeconomic terms, these strategies thereby support the substrategy of self-reliance.

The political implications of this are examined in the next section. First, the theoretical and methodological assumptions and consequences stemming from the investigation will be reviewed.

The survey had to struggle with a few problems in method and measurement that made it very difficult to compile data. Just calculating the area of a field--not to mention that of a farm unit which comprised several fields scattered about the countryside--was no easy task. The fields rarely had any clear geometric boundaries, their borders often disappearing into bush vegetation.

Once field size was figured, comparability presented another serious methodological obstacle, even if one excludes all other factors and compares crops only. The predominant practice of intercropping involves so many different variables--which are, in turn, differentiated into two seasons per year--that the possiblities for making a quantitative comparison are greatly restricted.

Finally, the information on inputs and outputs is often unreliable, making the calculations based upon it inexact, too. Accordingly, a crop budget, whose accurate calculation would require monocultural farming, can be drawn up for intercropped fields only by pretending that the complementary crops have never been planted. Farm budgets, however, which include crop groups anyway, are not all that difficult to calculate. Fortunately, the latter are far more important for the premises of this study.

The following theoretical implications can be drawn from the problems described above.

- The specification of types of peasant production units must be refined in the future. For this and other reasons, the most important task is to enlarge the data base greatly by including many more farms in the survey. The deviations from the mean will always be very large, but individual features can be assessed more accurately.
- It would also be reasonable to extend future studies over a period of two harvests (one year) so that information on yields and returns could be checked. This would, of course, greatly increase the cost of the research. Due to limitations on time and money, such a scope was not possible with the present survey.

## Increasing Yield through Environmentally Sound Agricultural Techniques

The increase in yield made possible by ecological, i.e., environmentally sound and energy saving agricultural techniques, results from the fact that they use the environment and nature more intensively as production factors and ecology more intensively as a means of production than do traditional or even chemotechnical methods of farming. The field experiments, which were conducted in cooperation with farmers, combined three main measures: erosion control, planting in contour rows, and intercropping.

The initial inputs of these measures could be documented while the owners' subsequent involvement (including the expenditure of labor and materials) during the vegetation period are no longer reflected. Figures for the yields could not be obtained, either, because of the short duration of the study. However, a detailed empirical input-output analysis or a micro-economic cost-return calculation is desperately needed for the ongoing evaluation of the production techniques used. Similarly, the package of methods itself must be supplemented by a variety of ecological techniques for soil care and pest control, with chemical products being used only on a limited basis for "therapeutic" purposes and not "prophylactically."

From this follows that larger scale experiments must be conducted and that these must be well documented and evaluated from the microeconomic perspective. The experimental fields should also be established under realistic conditions. In other words, they should enlist the help and supervision of the peasants involved and should not be run simply as socially and economically independent model farms that provide scientifically "value-free" but socioeconomically irrelevant pilot studies.

Beyond the matter of ecological and economic appropriateness, it is already apparent that the technology of ecofarming has also shown itself to be appropriate to the sociocultural conditions of the study area because it builds on traditional techniques, is at least provisionally accepted by the farmers, and contributes to the satisfaction of the needs articulated in the discussions.

It seems particularly important that these noncapital intensive techniques can be used to grow cash crops but can also be applied in subsistence production, which is perceived to be more important. They can thus provide higher income --measured in kind or in money--without sacrificing the aspect of minimizing risks.

## Completeness of the Elements?

What results has the case study on Tanzania provided for the concept of ecodevelopment? The concept focused on three elements--needs, self-reliance, and environmental compatibility--which were integrated into the case study and checked for their practicability. This case was based on a narrowly defined geographical region and rural population so that agriculture was the primary interest. The needs touched primarily labor and income; self-reliance was a matter of agricultural production, namely, subsistence production; and environmental compatibility related to yield increasing techniques.

In the last domain, the technological, lies a key to socioeconomic improvement. If it is also to have a long-lasting effect, then technology--by which is meant the totality of ecologically oriented agricultural production techniques (ecofarming) and socioeconomic interdependence--must show that it is ecologically, economically, and socially appropriate for the given conditions. In these criteria for appropriate technology, the elements of ecodevelopment are repeated in reverse order and on a different level of abstraction. Surprisingly, both concepts--the policy of ecodevelopment as a "style" for revitalizing economic development and the technology of ecofarming as the paradigm of appropriate agricultural technology--were designed independently and partially tested.

Nevertheless, they both converge, mutually complementing each other. In a virtually dialectic process the technique initially conceived independently is integrated as a means into the broader framework of policy, into the goals. This occurs through the realization of its limiting conditions:

- It must be ecologically appropriate, environmentally sound
- It must be economically feasible at the farm level; that is, one must be able to carry it out at the microlevel of subsistence farming and ultimately at the macrolevel with the available indigenous means on a risk-minimizing, low-input, independent basis

- It must be socially and culturally viable; that is, it must be accepted by the users and must correspond to the population's needs

It is natural that the technologically determined order exactly inverts the politically determined goals. Politically speaking, the satisfaction of needs has priority. It must be economically consistent and must not have long term destructive effects on the natural resources and the environment, the bases of the economic process and social welfare.

The empirical scrutiny of ecodevelopment's elements afforded by applying them to a concrete situation has brought to light the convergence of the theoretical micro and macro levels; it has shown the translatability of the one into the other when mediated by the implementation of the concrete measures. In addition to establishing the plausibility of the concepts of ecodevelopment and appropriate technology, this has also provided a very convincing argument for their theoretical integrity.

# CHAPTER IX

# POLITICAL IMPLICATIONS

What implications are there for development policy? Does the concept of ecodevelopment define a political path that can be followed? Which conclusions are valid for the development aid of the donor countries and organizations? What are the "worthwhile" aspects in the interests of the recipient countries?

## Needs Orientation

For all its theoretical shortcomings and all the methodological inadequacies involved in conducting it, the study of needs showed two things relevant for the evaluation of development policy:

- The discussion participants had very clear ideas about their needs and their current priorities
- The results from the need survey tallied with the objectively obtained data on material requirements

The two statements are by no means identical. For example, an interviewee could express a definite desire for products of western civilization like portable radios, pocket calculators, and cosmetics (which undoubtedly play a role already), thereby contradicting the "objective" data on material requirements.

Therein arises the problematic nature of the political significance and planning relevance of articulated needs. The importance of assessing needs and considering them part of the planning conditions is beyond question if one does not wish to assume the risk of ignoring the population during the planning process. Of course, this study was only an initial step toward involving those affected in the foundation and execution of planning. It is precisely when contradictions between "subjective" and "objective" data arise that it becomes necessary to enlist the population's participation in the planning process in a way that resolves the contradictions. A planning procedure is a failure if it is rejected by the very population it was meant to serve. It is an indication that the allocation of resources, which are usually scarce, was improperly managed.

Participation, on the other hand, has the advantage of minimizing the risk of failure by subjecting each stage of planning to review. Eliminating contradictions means that conflicts stemming from different interpretations of needs are recognized early, when it is still possible to balance competing interests against each other.

Naturally, such a concept can be realized more efficiently if the recipient country itself assumes responsibility for the planning and the donor country integrates its technical assistance and expertise into that process. Judging from the creation of so-called integrated development projects, it is possible that this call has largely been accepted since the mid-1970s.

A note of caution must still be sounded, however, as the impression has grown in the developing countries that the need concept has been used by industrialized countries to cheat the recipients out of the aid necessary for industrialization (see, for example, the Provisional Report, 1979, p. 11). An important political consequence is the necessity to emphasize that this fear is ungrounded since the planning authority as described here must lie with the recipients; those directly involved must participate in the process.

## Self-Reliance through Decentralized Production

Peasant agricultural production has most of the responsibility for supplying the country with food, and this for the most part with traditional farming methods using family labor and knowledge handed down from earlier generations. External inputs are the exception, not the rule. It is not a high-yield, efficient agricultural system, but because of its many small and decentralized production units, it is almost impervious to crises. Accordingly, this study's results showed that ensuring subsistence and distributing the risks have high priority over taking advantage of profit-making opportunities by, say, intensifying cultivation of export crops that are dependent on world markets and prices.

An optimal strategy would thus be one that introduces measures to increase yields (see the following section) by building on traditional knowledge and by preserving and enhancing the intact social structures upon which the peasant production structures and production decisions are based.

Admittedly, a high degree of subsistence supply does not stimulate the modern economic process, but it reduces the economic system's vulnerability to crises and thus contributes to political stability. The same is true for the high level of agricultural employment--including self-employment--that is achieved in this way. If one also considers that self supply involves not only the family subsistence and neighborly barter economy in the narrow sense but also the supply to the local markets, then it becomes clear that domestic economic impulses can be spurred in this way with indigenous resources and without additional external inputs via the consumption cycle.

This should not be confused with total autarchy and abstinence from the world markets. The special preconditions of the Chinese model, of coping with an entire period of development while relying exclusively on indigenous resources, do not exist elsewhere. But it should be emphasized that the subsistence domain in the broader sense must not be neglected in favor of total reliance on foreign trade. This domain, whose independence from external economic influences makes it the backbone of the national economy, should instead be nurtured and modernized in the interest of achieving greater productivity.

## Environmental Compatibility of Measures to Increase Yields

The modernization of agriculture for the purpose of achieving greater productivity should promote the capacity to rely on one's own potential and resources and should preserve and gradually improve the foundation of the agricultural system--the soil. Environmentally sound measures to increase yield are therefore an absolute necessity and represent the third element of ecodevelopment policy.

The enviornmentally compatible techniques of ecological farming are designed primarily to control erosion and to preserve soil fertility without using foreign material resources, by introducing "only" ecological information on geochemical cycles. It is the knowledge about these cycles that makes it possible to exploit additional resources when the agricultural system as an agro-ecosystem conforms as much as possible to the principles of a natural ecosystem.

The environment is promoted to the rank of a production factor here. The environmental compatibility of measures designed to increase yield contributes to preserving and improving the productive foundations of agriculture--the soil's resources, which can be renewed only with great difficulty. This, in turn, helps make the population's food supply less vulnerable to crises. Long range policy planning should thus aim to promote appropriate basic research and its application in large-scale experiments.

These kinds of measures belong in the catalogue of strategies for saving energy, protecting nature and the environment, and conserving resources. The energy argument even pertains indirectly to the balance of foreign exchange payments since the chemotechnical agricultural methods require five to ten times more energy than is returned to their yields (measured in energy units, excluding solar energy input). Mechanization, fuel consumption, and fertilizer comprise 79 percent of this figure (Pimentel et al., 1973,

p. 445, table 2). This means that these methods are heavily dependent on oil. This link can be weakened, however, by the greatest possible use of energy-saving, ecological techniques. A far sighted supply program, a long range economic policy, and a judicious policy of resource conservation must therefore embrace the environmentally sound production techniques of ecological farming.

## The Significance for Development, Aid Strategies, and Measures

What are the implications for the feasibility and applicability of the concept of ecodevelopment? Above all, under what conditions and to which degree does ecodevelopment provide a viable path for development policy?

The elements of the ecodevelopment concept--need satisfaction, self-reliance, and environmental compatibility --are taken from the political domain. The challenges they present have to do more with practical policy than with theory. Their political significance has been described and is obvious. Clearly, the practical relevance of these elements is demonstrated in detail by applying them--this study reflects that well, too--not just by depicting them in a wider context. In reality, they are intertwined and difficult to isolate from one another.

This can be exemplified by the satisfaction of needs. Planning that focuses on needs functions as a politically stabilizing element. Every deviation incurs costs that can range from reduced labor productivity and increased propaganda costs to, in extreme cases, repression. The problem lies in the details--in the question of the articulated and especially of the unarticulated needs and in the assessment of their political importance. Needs do not become relevant at the time and place a person is asked about them. They become relevant when they are acted upon.[1] This is the case especially when daily reproduction, work, and field preparation are concerned. Given production conditions and the means of production, the production strategy selected--which crops to grow in which combination and how intensively--reveals relevant and observable information on needs. Each improvement of the production conditions increases the level of satisfaction independently of qualitative changes in the need spectrum. A requisite of all recommendations for improvement is that they be economically feasible on the basis of indigenous resources (self-reliance at the microlevel) at an unchanging level of operational or market risks and that they safeguard long-term soil fertility, which rules out ruinous exploitation of the limited production factor of land (environmental compatibility).

This "microstrategic" perspective has implications for the macroeconomic and macropolitical decision-making level. The action model of ecodevelopment disencumbers policy, the economy, and the environment--

*policy* by encouraging decentralization and self-determination at the production level, both of which function as

---

[1]This is also a cognitive matter that has not yet yet been adequately treated by epistemology.

stabilizing elements;
*the economy* by making optimal use of human and natural resources, which, at the international level, improves the balance of trade and the balance of payments and reduces the dependence on world markets and, on the domestic level, creates employment and a condition for investment by demand;
*the environment* by conserving limited natural resources, especially by preserving the supply of fertile farmland.

This brings up not only the question of possibilities for the ecological path toward development, but also of its limits, its conditions and feasibility. One could summarize by saying that the possibilities lie in the political and economic latitudes of action that emerge. In general, national capital and foreign exchange are freed for additional development efforts, particularly for industrialization and the development of an infrastructure, perhaps also for rural and urban social improvements to the extent that the yields from the investments are sufficient for the debt service. Politically, internal capacity for action develops for reforms and legislation that can effect a possible redistribution of land or income. External capacity for action develops for the expansion of positions in foreign and trade policy such as the issue of the International Monetary Fund, the agreements on raw materials, or development aid accords.

Ecodevelopment must not be misinterpreted as a politicoeconomic panacea. Implementing the measures is tedious precisely because the details play such a great role. A certain overemphasis on economic and political demands is necessary because they have been neglected all too often in the past. But this cannot mean that total economic autarchy, for example, would be a worthwhile goal or that advantages of location that can be exploited through foreign trade should be ignored. Still, the basic premise should hold that available foreign exchange be spent on measures that will increase productivity and promote development rather than on luxury goods for an elite. A further limitation is that ecodevelopment can definitely be supplemented by measures in the domain of foreign trade. The recommendation of reforming the world economic order to give developing countries a range of advantages and support in foreign trade does take a completely different approach toward development, but its impact need not be thought to conflict with ecodevelopment.

The feasibility and political nature of ecodevelopment involves conflicting interests since, like every policy (especially development policy) it touches directly or indirectly on problems of distribution and provokes distribution conflicts. Ecodevelopment should benefit the poorest social strata most, particularly the rural ones. How much the politically better represented interest groups of the bureaucratic elites, the commercial and industrial middle class, the urban proletariat, and others are moved to cooperate or at least to forego certain things temporarily is contingent upon each group's political and economic latitude and how it is used for supporting measures. The economic interests of the trading partners and private industry and their demand for raw materials present international difficulties because the policy of ecodevelopment insists on massive import substitution, made possible by the mobilization of local resources.

If the partially conflicting statements and interests are melded, then ecodevelopment appears, in both political and economic terms, to be a path toward development that is definitely worth taking. That its implementation will not always be simple and that it can offend established interests must be accepted.

Donors who do not simply finance short-term exploitation of raw materials but who also have an interest in the long-term economic and political stability of their partners are advised to support ecodevelopment policies. That is not very easy, of course, since there are obstacles to promoting such policies directly. Ecodevelopment is designed in such a way that foreign capital aid need not, indeed must never, be used so that the obligation to pay it back will never arise. In order to arrive at "appropriate" and "need-oriented" solutions, technical aid must, in turn, focus so intensively on the details that it would be difficult to spend significant sums of money.

From this paradox of development policy, a solution can probably be found only through an intensive exchange of ideas with the recipients about direct and indirect measures that would be worth supporting. Sound development policy is difficult to shape because it demands a great deal of imagination, flexibility, understanding, and time.

# QUESTIONNAIRES

## General Household Inquiry[1]

1. Personal data

   1.1. Place of residence ______________________
   1.2. Name of father ______________________
   1.3. Name of family (clan) ______________________
   1.4. Name of tribe ______________________
   1.5. Religion of father ______________________
   1.6. Age of father ______________________

2. Family size

   2.1. Number of persons living in the household
   2.1.1. Father: alive ( ) dead ( ) ( )
   2.1.2. Number of wives ( )
   2.1.3. Number of sons ( )
   2.1.4. Number of daughters ( )
   2.1.5. Number and relationship of others ( )

   ................................................

   2.2 Number of children not living in the household ( )

3. Educational level of family members

   3.1. Father's education ______________________
   3.2. Mother's education ______________________
   3.3. Sons' education ______________________
   3.4. Daughters' education ______________________

4. Working status

   4.1. Employed ( ) Type of employment __________
   4.2. Pension
   4.3. Not employed ( ) Type of work __________

5. Agricultural production

   5.1. Size and number of *shambas* (fields) ______________
   5.2. Employing other persons yes ( ) no ( )

[1]Developed and conducted jointly with H.W. Schönmeier, University of Saarbrücken.

5.3. Type of agricultural production
5.3.1. Subsistence (maize, cassava, . . . ) yes ( ) no ( )
5.3.2. Cash crop vegetables yes ( ) no ( )
5.3.3. Cash crop coffee yes ( ) no ( )
5.3.4. Fruit trees (avocado, mango, . . . ) yes ( ) no ( )
5.3.5. Others yes ( ) no ( )

.............................................

5.4. Animal husbandry
5.4.1. Cattle yes ( ) no ( )
5.4.2. Sheep/goats yes ( ) no ( )
5.4.3. Pigs yes ( ) no ( )
5.4.4. Chickens yes ( ) no ( )
5.4.5. Others yes ( ) no ( )

.............................................

5.5 Application of fertilizer and pesticides
5.5.1. Fertilizer yes ( ) no ( )
5.5.2. Pesticides yes ( ) no ( )

6. Type of house owned by the family

6.1. House where father lives
6.1.1. Corrugated iron roof yes ( ) no ( )
6.1.2. Cement floor yes ( ) no ( )
6.1.3. Number of rooms number ( )
6.2. Other houses of the family number ( )
6.2.1. Corrugated iron roof yes ( ) no ( )
6.2.2. Cement floor yes ( ) no ( )

7. Communal activities

Do members of the family work together with others or do they join communal activities?
communal activities ( )
work with others ( )

What activities? ______________________________

## Basic Needs

Name of person interviewed ______________________ No. ___

1. Which of the following items are most important for you and your family? State in order of preference.

1.1. Food . . . . . . . . . . . . . . . . . . . . . . ( )
1.2. Clothing . . . . . . . . . . . . . . . . . . . . ( )
1.3. House . . . . . . . . . . . . . . . . . . . . . ( )
1.4. Health . . . . . . . . . . . . . . . . . . . . . ( )
1.5. Education . . . . . . . . . . . . . . . . . . . ( )
1.6. Farm land . . . . . . . . . . . . . . . . . . . ( )
1.7. Employment . . . . . . . . . . . . . . . . . . ( )

1.8. Transport to your *shamba* (field), to the market or hospital . . . . . . . . . . ( )
1.9. Travelling to Mombo, Tanga, or Arusha . . . . ( )
1.10. Free time without work . . . . . . . . . . . ( )

2. Why did you place a check next to these and not others? Please explain ______________________________

______________________________

______________________________

3. If you had the choice, which type of work would you like to do? Would you like to be a

3.1. Farmer . . . . . . . . . . . . . . . . . . . . . ( )
3.2. Worker . . . . . . . . . . . . . . . . . . . . . ( )
3.3. Tailor . . . . . . . . . . . . . . . . . . . . . ( )
3.4. Middleman . . . . . . . . . . . . . . . . . . . . ( )
3.5. Teacher . . . . . . . . . . . . . . . . . . . . . ( )
3.6. Politician . . . . . . . . . . . . . . . . . . . ( )
3.7. Any other . . . . . . . . . . . . . . . . . . . . ( )

Please respond to each question or statement below.

4. What is the reason for your choice in comparison with other alternatives? ______________________________

5. What work do you actually do? ______________________________

______________________________

6. If different, why don't you work in the profession that you like? ______________________________

______________________________

7. Which of the following reasons is also important for the choice of the profession you like?

| | | Yes | No |
|---|---|---|---|
| 7.1. | I have more food . . . . . . . . . . . . . | ( ) | ( ) |
| 7.2. | I earn more money . . . . . . . . . . . . . | ( ) | ( ) |
| 7.3. | I like to work for the government . . . . . | ( ) | ( ) |
| 7.4. | I like a regular income . . . . . . . . . . | ( ) | ( ) |
| 7.5. | I want to save money in the bank . . . . . | ( ) | ( ) |
| 7.6. | When I am old, I will get a pension . . . . | ( ) | ( ) |
| 7.7. | I want to work in a group with people I like | ( ) | ( ) |
| 7.8. | I am a big man; people will admire me | ( ) | ( ) |
| 7.9. | This is just the type of work that I love to do. It makes me feel happy . . . . . . | ( ) | ( ) |
| 7.10. | It is best for my family . . . . . . . . . | ( ) | ( ) |
| 7.11. | It pleases the people in my village . . . . | ( ) | ( ) |

Please answer each question.

8. If the answer is "more money," what do you need it for? ______________________________

______________________________

9. Labor time

9.1. How many hours do you work daily? ______________________
9.2. How many days per week? ______________________
9.3. Does this number change during the year? yes ( ) no ( )
By how much? ______________________

10. During your free time, what do you prefer to do?

______________________

11. What else do you like to do during your free time?

| | | Yes | No |
|---|---|---|---|
| 11.1. | I eat food or drink *pombe* (local beer) . . . | ( ) | ( ) |
| 11.2. | I talk with friends . . . . . . . . . . . . . | ( ) | ( ) |
| 11.3. | I stay with my family . . . . . . . . . . . . | ( ) | ( ) |
| 11.4. | I prefer to go farming . . . . . . . . . . . . | ( ) | ( ) |
| 11.5. | I work for money . . . . . . . . . . . . . . . | ( ) | ( ) |
| 11.6. | I work for the party or the government . . . | ( ) | ( ) |
| 11.7. | I go to school, adult education, or vocational training . . . . . . . . . . . . | ( ) | ( ) |

Please answer each question.

12. Some people say that one could grow more crops and earn more money by working harder.

12.1. Is it possible to work harder? . . . . . . . . ( ) ( )
12.2. Is it more important to have free time or to work harder? Why? ______________________

______________________

13. There is a shortage of land in Shashui.

13.1. How many members of your family went elsewhere to farm? ______________________
13.2. Are there people in your village who do not own own land? . . . . . . . . . . . . . . . . . . ( ) ( )

14. Since farm land is scarce, some people say it would be easier to feed a small family than a large one.

14.1. Do you think most people should have fewer children? . . . . . . . . . . . . . . . . . . ( ) ( )
14.2. Do you personally plan to limit the number of your children? . . . . . . . . . . . . . ( ) ( )
14.3. If yes, for which reason? . . . . . . . . . . ( ) ( )
14.4. How many children would you like? ______________________

15. Resettlement

| | | Yes | No |
|---|---|---|---|
| 15.1. | Are there any resettlement plans in the district so that people are asked to move into bigger villages? . . . . . . . . . . . | ( ) | ( ) |

*[Interviewer: If yes, ask the following questions.]*

15.2. According to your opinion, what are the reasons?

________________________________________

15.3. What are the advantages? ____________________

________________________________________

15.4. What are the disadvantages? __________________

________________________________________

## Inputs and Outputs

Name of person interviewed ____________________ No. ___

1. How many *shambas* (fields) do you have? ________________

2. Where are these *shambas* located and how do you use them?

| | Location | Crops | Acres |
|---|---|---|---|
| First *shamba* | | | |
| Second *shamba* | | | |
| Third *shamba* | | | |
| Fourth *shamba* | | | |
| Fifth *shamba* | | | |
| Sixth *shamba* | | | |
| Seventh *shamba* | | | |
| Eighth *shamba* | | | |
| Ninth *shamba* | | | |
| Tenth *shamba* | | | |

3. What is the size of all your *shambas* in acres? __________

4. Besides these *shambas*, do you have additional land?

yes ( ) no ( )

How much? ________________ What type of land? __________

Where is it located? ______________________________

Why is it not used as a *shamba* at present? ______________

________________________________________

5. Which yields did you get last year (two rainy seasons)?

| Crop | Shamba 1st | 2nd | 3rd | 4th | 5th | 6th | 7th | Total Yield (in kg) |
|---|---|---|---|---|---|---|---|---|
| Coffee | ___ | ___ | ___ | ___ | ___ | ___ | ___ | ______ |
| Tea | ___ | ___ | ___ | ___ | ___ | ___ | ___ | ______ |
| Vegetables | ___ | ___ | ___ | ___ | ___ | ___ | ___ | ______ |
| ______ | ___ | ___ | ___ | ___ | ___ | ___ | ___ | ______ |
| ______ | ___ | ___ | ___ | ___ | ___ | ___ | ___ | ______ |
| ______ | ___ | ___ | ___ | ___ | ___ | ___ | ___ | ______ |
| Beans | ___ | ___ | ___ | ___ | ___ | ___ | ___ | ______ |
| Maize | ___ | ___ | ___ | ___ | ___ | ___ | ___ | ______ |
| Cassava | ___ | ___ | ___ | ___ | ___ | ___ | ___ | ______ |
| Bananas | ___ | ___ | ___ | ___ | ___ | ___ | ___ | ______ |
| ______ | ___ | ___ | ___ | ___ | ___ | ___ | ___ | ______ |
| ______ | ___ | ___ | ___ | ___ | ___ | ___ | ___ | ______ |
| ______ | ___ | ___ | ___ | ___ | ___ | ___ | ___ | ______ |
| ______ | ___ | ___ | ___ | ___ | ___ | ___ | ___ | ______ |

6. If you had one more field of 1 acre size, what would you plant? (in detail)

Crops ______________________________

Quantity ______________________________

7. What is the reason for your decision? ______________

______________________________

______________________________

8. If you had to give 1 acre away, which crops would you stop growing? (in detail)

Crops ______________________________

Quantity ______________________________

9. What is the reason for your decision? ______________

______________________________

10. What crops are the most profitable? ______________

______________________________

11. Why don't you grow more of these profitable crops?

Shortage of labor ( ); Prefer food crops ( );
Shortage of money ( ); Lack of interest ( );

Other reasons:

______________________________

## Production Increase

Name of person interviewed ______________________ No. ___

1. Which crops do you grow? ______________________

______________________

2. What are your main farming or agricultural problems?

2.1. Problem 1 ______________________
2.2. Problem 2 ______________________
2.3. Problem 3 ______________________
2.4. Problem 4 ______________________
2.5. Problem 5 ______________________

3. Can you do anything about that? . . . . yes ( ) no ( )

4. If yes, what have you done to solve your problems?

4.1. Problem 1 ______________________
4.2. Problem 2 ______________________
4.3. Problem 3 ______________________
4.4. Problem 4 ______________________
4.5. Problem 5 ______________________

5. Did you or do you receive help to solve your problems? . . . . . . . . . . . . . yes ( ) no ( )

5.1. If yes, who helped (helps) you?:

5.1.1. Family, friends, neighbors? . . . . yes ( ) no ( )
5.1.2. Kilimo, LIDEP? . . . . . . . . . . yes ( ) no ( )
5.1.3. Any others? (specify) ______________________

5.2. What kind of help do you get? ______________________
5.3. Even if nobody has helpd you, would anyone be *able* to solve your agricultural problems? ______________________

______________________

6. Do you want to have an increase in production or are you happy with the yields you have? Increase ( ) Happy ( )

6.1. Please specify the reason for your answer: ______________________

______________________

6.2. Is it even possible to increase yields on your *shambas*?

yes ( ) no ( )

6.3. Please specify the reason for your answer: ______________________

______________________

7. Which of the following methods would you consider useful in order to increase your production? (Please respond to each possibility.)

| | | Yes | No |
|---|---|---|---|
| 7.1. | Use more labor | | |
| 7.1.1. | Work longer . . . . . . . . . . . . . . . . | ( ) | ( ) |
| 7.1.2. | Employ other people . . . . . . . . . . . . | ( ) | ( ) |
| 7.2. | Use more or better material inputs | | |
| 7.2.1. | Better seedlings . . . . . . . . . . . . | ( ) | ( ) |
| 7.2.2. | More animal or plant manure . . . . . . . | ( ) | ( ) |
| 7.2.3. | Fertilizer . . . . . . . . . . . . . . . | ( ) | ( ) |
| 7.2.4. | Pesticides . . . . . . . . . . . . . . . | ( ) | ( ) |
| 7.3. | Use mechanized tools | | |
| 7.3.1. | Plow . . . . . . . . . . . . . . . . . . | ( ) | ( ) |
| 7.3.2. | Harrow . . . . . . . . . . . . . . . . . | ( ) | ( ) |
| 7.3.3. | Cultivator . . . . . . . . . . . . . . . | ( ) | ( ) |
| 7.3.4. | Oxen for drawing . . . . . . . . . . . . | ( ) | ( ) |
| 7.3.5. | Tractors for drawing . . . . . . . . . . | ( ) | ( ) |
| 7.4. | Irrigate | | |
| 7.4.1. | Carry water . . . . . . . . . . . . . . . | ( ) | ( ) |
| 7.4.2. | Dig an irrigation trench . . . . . . . . | ( ) | ( ) |
| 7.5. | Use ecological methods | | |
| 7.5.1. | Shade trees on the *shamba* . . . . . . . . | ( ) | ( ) |
| 7.5.2. | Horizontal ridges of Guatemala grass, trees, etc. . . . . . . . . . . . . . . . | ( ) | ( ) |
| 7.5.3. | Intercropping . . . . . . . . . . . . . . | ( ) | ( ) |
| 7.5.4. | Crop rotation . . . . . . . . . . . . . . | ( ) | ( ) |
| 7.5.5. | Mulching instead of burning dry material | ( ) | ( ) |
| 7.5.6. | Reduced weeding instead of clean weeding | ( ) | ( ) |
| 7.5.7. | Leave fields fallow . . . . . . . . . . . | ( ) | ( ) |

8. According to your experience, what are the best agricultural methods? ______________________________

______________________________________________

______________________________________________

9. Why are these methods the best? ____________________

______________________________________________

10. Is this true for all crops you grow? What differences are there? ______________________________

______________________________________________

## SELECTED BIBLIOGRAPHY

Acland, J.D. East African Crops. London: Paul Longman Group in arrangement with the Food and Agriculture Organization of the United Nations, 1971.

Becker, H. and Kiefer, W. "Weinbau und Kellerwirtschaft in der Region Dodoma, Tanzania." Paper prepared for the Kübel Foundation, Bensheim, 1977. (Mimeographed.)

Brandt, H.; Gerken, E.; and Schubert, B. The Industrial Town as a Factor of Economic and Social Development: The Example of Jinja, Uganda. IFO Series on African Studies, no. 77. Munich: Weltforum Publishing House, 1972.

Brandt, H. "Zur Steigerung der Beschäftigung in afrikanischen Landgebieten - das Arbeitsangebot." Zeitschrift für ausländische Landwirtschaft 13 (1974): 97-110.

Brede, K. and Siebel, W. "Zur Kritik der Bedürfnisforschung." Leviathan 1 (1977): 1-27.

Bretano, L. "Versuch einer Theorie der Bedürfnisse." Proceedings of the Royal Bavarian Academy of Sciences, Philosophic, Philologic, and Historical Class, Tenth Treatise. Munich: Königliche Bayerische Akademie der Wissenschaften und G. Franz'schen Verlag (J. Roth), 1908.

Clayton, E.S. "Agrarian Reform, Agricultural Planning and Employment in Kenya." International Labour Review 102 (1970): 431-53.

Egger, K. "Ökologie als Produktivkraft - Erfahrungen bei 'Ecofarming' in Ostafrika." In Agrarreform in der Dritten Welt, pp. 217-54. Edited by H. Elsenhans. New York: Campus, 1979.

Egger, K. and Glaeser, B. Politische Ökologie der Usambara-Berge in Tanzania. Kübel Foundation Series. Bensheim: Kübel Foundation, 1975.

Egger, K. and Mayer, B.J. "Agro-Technological Alternatives for Agriculture in the Usambara Mountains - Ecofarming." Institute for Botany, University of Heidelberg, 1976. (Mimeographed.)

Elwert, G. and Wong, D. "Structures Agraires et procès de développement dans la province de l'atlantique. Department of Sociology, University of Bielefeld, 1978. (Mimeographed.)

Feder, G. and O'Mara, G.T. "Farm Size and the Diffusion of the Green Revolution Technology." Economic Development and Cultural Change 30 (October, 1981): 59-76.

Fleuret, A. "Physical Mobility Among Shambaa Women." [1976].

Fleuret, A. and Fleuret, P. "Firewood Report." [1976].

Fleuret, P. "Produce in Northeast Tanzania." [1976].

German Agricultural Team (GAT). Growing Potatoes in Kenya. Edited by the Federal Agency for Economic Cooperation (B.f.E.), Division of Agriculture. Essen: Webels, 1974.

Glaeser, B. "Sozialökonomische Bestandsaufnahme und Problematik in den Usambara-Bergen." In Politische Ökologie der Usambara-Berge in Tanzania, pp. 139-222. Kübel Foundation Series. Bensheim: Kübel Foundation, 1975.

Glaeser, B., ed. Vierteljahresberichte - Probleme der Entwicklungsländer, no. 70, special issue: "Ecodevelopment - Aspects and Problems of an Alternative Rural Development" (1977).

Glaeser, B. "Economic Implications of Ecological Sustainability in the Agricultural Development of East Africa." Vierteljahresberichte - Probleme der Entwicklungsländer, no. 70, special issue: "Ecodevelopment - Aspects and Problems of an Alternative Rural Development" (1977): 305-16.

Glaeser, B. "'Ecodevelopment' und seine Auswirkungen auf den Energieverbrauch." In Energieversorgung und Lebensqualität, pp. 358-76. Edited by V. Hauff. Series on "Argumente in der Energiediskussion," vol. 6. Villingen: Neckar, 1978.

Grimwood, B.E. The Processing of Macadamia Nuts. London: Tropical Products Institute, 1971.

Hekstra, G.P. "Towards a Conservation Strategy to Retain World Food and Biosphere Options." In Food-Climate Interactions, pp. 325-59. Edited by W. Bach, J. Pankrath, and S.H. Schneider. Dordrecht, Holland: D. Reidel, 1981.

Herrera, A.O.; Scolnik, H.D.: Chichilnisky, G.; Gallopin, G.C.; Hardoy, J.E.; Mosovich, D.; Oteiza, E.; de Romero Brest, G.L.; Suárez, C.E.; and Talvera, L. Catastrophe or New Society? A Latin American World Model. Ottawa: International Development Research Centre, 1976.

Hozza, J.J. "The Hoza Rebellion and After: A Study in Innovation." Ph.D. dissertation, Political Science Paper 6, The University College, Dar es Salaam, 1976.

Kimela, K.G. "A Survey on Ecologically Based Advice for Farming in Five Primary Schools in Lushoto District - Preliminary Report." Paper prepared for the Lushoto Integrated Development Project (LIDEP), Soni, 1976. (Mimeographed.)

Lederer, K. "Bedürfnisforschung und Umweltplanung." In Proceedings of the Second Vienna International Meeting on Human Ecology, pp. 117-26. Edited by H. Knötig. Vienna: Archivum Oecologiae Hominis, 1977.

Lederer, K., ed. Human Needs--A Contribution to the Current Debate. Cambridge, Mass.: Oelgeschlager, Gunn & Hain, 1980.

Lederer, K. and Mackensen, R. Gesellschaftliche Bedürfnislagen. Series of the Commission for Economic and Social Change, no. 92. Göttingen: Otto Schwarz & Co., 1975.

Lenk, H. "Zu neueren Ansätzen der Technikphilosophie." In Techne-Technik-Technologie, pp. 198-231. Edited by H. Lenk and S. Moser. UTB Paperback Books, no. 289. Munich: Saur Verlag, 1973.

Lockeretz, W.; Klepper, R.; Commoner, B.; Gertler, M.; Fast, S.; O'Leary, D.; and Blobaum, R. "A Comparison of the Production, Economic Returns, and Energy Intensiveness of Corn Belt Farms That Do and Do Not Use Inorganic Fertilizers and Pesticides." Washington University, Saint Louis, Mo.: CBNS-Ae 4, 1975.

Macpherson, G.A. First Steps in Village Mechanisation. Dar es Salaam: Tanzania Publishing House, 1975.

Macpherson, G.A. and Jackson, D. "Village Technology for Rural Development--Agricultural Innovation in Tanzania." International Labour Review 111 (1975): 97-118.

Malinowski, B. A Scientific Theory of Culture. Chapel Hill, North Carolina: University of North Carolina Press, 1944.

Maslow, A.H. Motivation and Personality. New York: Harper & Row, 1954; 2nd ed., New York: Harper & Row, 1970, pp. 35-58.

Meyer, P. "Vegetable Growing in Lushoto District." Paper prepared for the Lushoto Integrated Development Project (LIDEP). Soni, 1971. (Mimeographed.)

Moore, J.E. "Rural Population Carrying Capacities of the Districts of Tanzania." Bureau for Research Assessment and Land Use Planning (BRALUP), Research Paper No. 18. University of Dar es Salaam, 1971.

Newcombe, K. "Energy Use in the Hong Kong Food System." Agro-Ecosystems 2 (1975): 253-76.

Pimentel, D.; Hard, L.E.; Bellotti, A.C.; Forster, M.J.; Oka, I.N.; Sholes, O.D.; and Whitman, R.J. "Food Production and the Energy Crisis." Science 182 (1973): 443-9.

Pretzell, M. "Die Welternährungslage 1977." Entwicklungspolitik. Materialien Nr. 59. Bonn: Federal Ministry for Economic Cooperation, 1977.

"Provisional Report on the Conference Deliberations at the Second International Conference on Perspectives in Afro-German Relations." Arusha, United Republic of Tanzania, 8-12 January 1979. (Mimeographed.)

Schönmeier, H.W. Agriculture in Conflict--The Shambaa Case. Kübel Foundation Series. Bensheim: Kübel Foundation, 1977.

Schoess, C. "Hier haben wir noch Motivationsreserven!" Management Zeitschrift Industrielle Organisation, special issue: "Zukunft der Marktwirtschaft - Marktwirtschaft der Zukunft" (1977): 77-80.

Steinhart, J.S. and Steinhart, C.E. "Energy Use in the U.S. Food System." Science 184 (1974): 307-16.

Tschiersch, J.E. Angepaßte Formen der Mechanisierung bäuerlicher Betriebe in Entwicklungsländern. Heidelberg: Forschungsstelle für internationale Agrarentwicklung, 1975.

The United Republic of Tanzania, Regional Development Director's Office. Tanga Regional Development Plan 1975-1980. Tanga: Tanga Integrated Rural Development Programme (TIRDEP), 1975.

What Now? - Another Development. The Dag Hammerskjöld Report, Development Dialogue, no. 1/2 (1975).